Now, let's go into "The World of Plants" of the National Museum of Natural History and find the green life on this blue Earth.

THE GREEN OF LIFE

生命之绿

马莉 编著

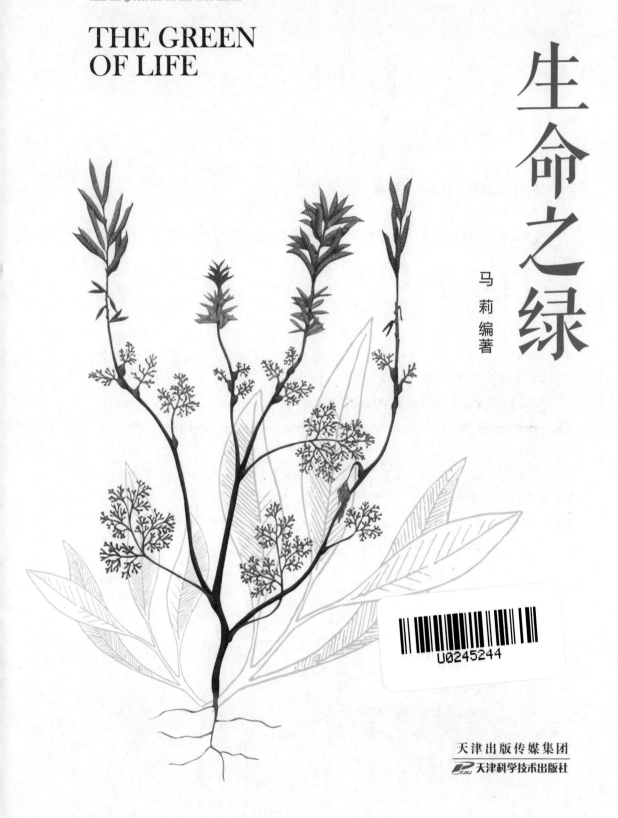

天津出版传媒集团

天津科学技术出版社

图书在版编目（CIP）数据

生命之绿 / 马莉编著. -- 天津 : 天津科学技术出版社，2023.12
 ISBN 978-7-5742-1661-7

Ⅰ. ①生… Ⅱ. ①马… Ⅲ. ①植物-青少年读物 Ⅳ. ①Q94-49

中国国家版本馆 CIP 数据核字(2023)第 208312 号

生命之绿
SHENGMING ZHI LÜ
责任编辑：韩　瑞
责任印制：兰　毅

出　　版：	天津出版传媒集团 天津科学技术出版社
地　　址：	天津市西康路 35 号
邮　　编：	300051
电　　话：	(022) 23332390
网　　址：	www.tjkjcbs.com.cn
发　　行：	新华书店经销
印　　刷：	天津印艺通制版印刷股份有限公司

开本 787×1092　1/16　印张 7.625　字数 57 000
2023 年 12 月第 1 版第 1 次印刷
定价：100.00 元

Contents
目 录

Chapter 1　The Discourse of Green Life
第1章　绿色生命的倾诉 ········· 1

Introduction of the National Museum of Natural History ········· 2
The Discourse of Green Life ········· 7
The World of Plants ········· 7
The Evolution of Plants ········· 7
Plants and Human Beings ········· 18
Diversification and Adaptation of Angiosperms ········· 23
走进国家自然博物馆 ········· 32
绿色生命的倾诉 ········· 34
植物演化 ········· 34
植物与人类 ········· 39
被子植物的繁盛与适应 ········· 41

Chapter 2　Getting Closer to Green Life Scientists
第2章　走近绿色生命的科学家 ········· 45

The Scientist Who Approached the Green of Life ········· 46
Carl Linnaeus, the Father of Biological Taxonomy ········· 46
Chinese Pioneer in Plant Taxonomy——Hu Xiansu ········· 50
The "Living Dictionary" of Chinese Plants——Wu Zhengyi ········· 53
The "Father of Hybrid Rice"——Yuan Longping ········· 57
The Power of a Small Grass——Tu Youyou ········· 60
The Botanist Next Door——Chen Shaoxing ········· 63
走近绿色生命的科学家 ········· 66
生物分类学之父——林奈 ········· 66
中国植物分类学奠基人——胡先骕 ········· 69
中国植物的"活词典"——吴征镒 ········· 71
"杂交水稻之父"——袁隆平 ········· 73

一株小草的力量——屠呦呦 …………………………………… 76

身边的植物学家——陈绍煋 …………………………………… 78

Chapter 3　The Beauty of Green Life
第3章　绿色生命之美 ……………………………………… 80

一棵苹果树 ……………………………………………………… 82

植物的演化历程 ………………………………………………… 84

如何划定低等植物与高等植物 ………………………………… 87

植物界的"猛兽家族" …………………………………………… 88

植物界的"大熊猫" ……………………………………………… 90

恐龙的见证者 …………………………………………………… 92

天安门的大花篮 ………………………………………………… 94

中国标本馆的前世今生 ………………………………………… 96

标本制作——化绿色生命之美为永恒 ………………………… 98

Chapter 4　Plants in Chinese Culture
第4章　中国文化中的植物 ……………………………… 102

出淤泥而不染,濯清涟而不妖 ………………………………… 105

可使食无肉,不可使居无竹 …………………………………… 107

唯有牡丹真国色,花开时节动京城 …………………………… 109

唯有此花开不厌,一年长占四时春 …………………………… 112

作者简介

马 莉
国家自然博物馆 副研究馆员
专业领域：青少年校外科普教育研究

开创国内博物馆第一家"小小科普讲解员培训"、"博物馆戏剧表演"等多项青少年科普教育课程，形成自然类博物馆的文化品牌；撰写原创科普剧剧本十余部，多次指导学生参加全国科普剧及科学表演大赛、全国艺术大赛及讲解比赛，并屡获大奖；出版《童语自然》《生生不息——自然博物馆里的科普剧》《解读自然：小小科普讲解员培训教材》《丈量演化的脚步》等图书，发表多篇科教论文。

卷首语

习近平总书记十分重视博物馆工作，曾强调"一个博物院就是一所大学校"。博物馆是保护和传承人类文明的重要场所，是联结过去、现在、未来的一道桥梁。国家自然博物馆秉承以文化服务于全社会的宗旨，充分发掘厚重馆藏和优质科普资源，紧抓"扣好人生第一粒扣子"，坚持开展"小小讲解员培养项目"的实践活动，自觉引导青少年主动学习自然科学知识，弘扬中华优秀传统文化以提高社会责任感，厚植家国情怀以增强历史使命感。这既是国家自然博物馆教育和科学传播功能的重塑和延伸，也是实现青少年科普素质提升与全面能力培养的有机融合。

自2003年始，国家自然博物馆"小小科普讲解员"培养项目至今已经整整持续了20年。正所谓，北海虽赊，扶摇可接。20年来博物馆累计培养小小科普讲解员7000余名，约数百万观众聆听了孩子们的讲解。他们不仅在各展厅从事讲解、导览、接待等日常公众服务，还深度参与国家自然博物馆的多项科普项目和活动。如央视、北京科教频道等多档科普类节目和自然博物馆科普宣传片的录制，以及多部原创科普剧的排练与展演，还有北京市科学表演大赛等。仅科普剧展演活动就累计演出逾百场，线上、线下受众人数多达50000余人。

本书选题源自于博物馆的基本展陈资源，多视角向同学们介绍展示植物世界的点滴知识。一改以往大多数同学来博物馆参观，更多会将关注度投入到以恐龙为主题的古生物学内容。而我们人类生活的地球，历经46亿年漫长的演化，直到4亿多年前因植物登陆才在陆地上呈现出这束微弱的生命之绿，但生命长河万古不息，从此开启了地上郁郁葱葱的繁华，并迎来了多彩生命世界的繁荣。可以说，是绿色植物滋养了地球万物。

"桃之夭夭，灼灼其华""彼泽之陂，有蒲与荷""桑之未落，其叶沃若"，从《诗经》考证，便知植物在中国古代传统文化中不可或缺的位置，诗词歌赋、小说话本，处处都有植物的身影。古代文人赋予了植物浓厚的人文气息，中医文化更是把植物功效发挥到极致。在近代科技史中，植物学名中竟然还蕴含着植物学家的生平、好恶，甚至折射出一个民族的气质，从植物信息中匡补历史缺憾、勾勒陈年旧事。从自然知识中品味丰富的传统文化，激发孩子们弘扬强国复兴有我的爱国主义精神。

时光回溯到 1951 年,国家自然博物馆的前身中央自然博物馆筹备伊始,筹备处 7 位科学家中就有 2 位是植物学家,其中胡先骕是中国植物分类学的奠基人,李继侗是植物生态学与植物学的奠基人。早在 1954 年,博物馆就开始举办以植物为主题内容的农产资源展,随后植物展就作为国家自然博物馆的四大基本陈列之一。

《生命之绿》一书不仅可以让孩子们辨识植物世界万千,也希望孩子们能走进那些立志探索绿色生命、默默奉献的科学家们的精神世界。在物质与技术相对匮乏落后的年代,老一辈科学家不畏艰难险阻,踏遍祖国的崇山峻岭,采集标本,整理分类,砥志研思。他们把毕生都献给了科研事业,才成就了中国植物学的发展在国际学术界勃勃生机的地位。在讲解中深度挖掘展品背后饱含的科学家的故事,希望孩子们在认知大自然的同时,更能感受到历代科研人员为了理想和信念,始终忘我地奋战在科研、科普一线的高尚情操与理想境界。

从首届"小小科普讲解员"培训班开班,到长期专注于青少年校外科普教育培训工作,我在这条充满苦乐酸甜的道路上已走过 20 余年。已不记得在科普教育过程中有多少回和孩子们分享妙趣横生的科学探索活动,有多少次在展厅孜孜不倦地传播科学知识、在录音棚全神贯注地录制科学故事、在舞台上酣畅淋漓地演绎动物传奇……律回岁晚冰霜少,春到人间草木知,是热爱与传承释放的力量,促使我能够持之以恒坚持下来并继续启程。

在此,我要感谢国家自然博物馆副馆长张玉光博士百忙之中对本书内容的提点与指导;感谢科普教育部主任赵洪涛博士多年来对于培训工作的支持与帮助;感谢 Maggie 老师对图书中英文部分的细致校对;感谢父母家人对我的默默支持,特别是我的父亲马连成先生,我能够心无旁骛的投入工作,离不开他们多年无私的付出。

囿于自己水平有限,书中难免存在错漏之处,恳请读者及同行专家批评指正。

马 莉

2023 年仲秋

第 1 章　绿色生命的倾诉

Chapter 1　The Discourse of Green Life

Introduction of the National Museum of Natural History

The National Museum of Natural History is the only national and comprehensive natural museum in China, representing the national protection, research, collection, interpretation, and display of natural heritage with historical, scientific, and artistic value in the process of natural and human development.

Dear students, whenever we stroll through the exhibition halls of the National Museum of Natural History, stop at the dazzling exhibits, and marvel at the magic of biological evolution, have we ever thought that while understanding the historical changes of the earth, we also understand the development history of the National Museum of Natural History?

Now, we are at the central axis of the southern city of Beijing, the capital, to see the first large-scale Natural History Museum built by the People's Republic of China.

Its predecessor was the Preparatory Office of the Central Natural History Museum established on April 2, 1951. At that time, there were seven scientists and five leaders of the Ministry of Culture in the Preparatory Office.

中央自然博物馆筹备委员会第一次会议
前排从左至右：裴文中、郑作新、胡先骕、丁西林、李继侗
后排从左至右：王冶秋、张春霖、刘钧、李璞

第 1 章　绿色生命的倾诉
Chapter 1　The Discourse of Green Life

The seven scientists are as follows:

Hu Xiansu——a member (academician) of the Chinese Academy of Sciences, a founder of Chinese plant taxonomy.

Zheng Zuoxin——a member of the Chinese Academy of Sciences (academician), is the founder of modern Chinese ornithology and the pioneer of zoogeography.

Li Jitong——a member (academician) of the Chinese Academy of Sciences, a pioneer of plant physiology in China, and one of the founders of plant ecology and botany.

Gao Zhenxi——a member (academician) of the Chinese Academy of Sciences, is the main founder of the China Geological Museum and China's first national geological nature reserve.

Sun Yunzhu——a member of the Chinese Academy of Sciences (academician), the founder of Chinese paleontology, and one of the founders of the Chinese Geological Society.

Professor Li Pu——a famous geologist and the pioneer of isotope geochemistry in China.

Professor Zhang Chunlin——a founder of modern Chinese ichthyology.

The five leaders of the Ministry of Culture are all well-versed in both Chinese and Western cultures, as well as arts and sciences:

Ding Xilin——a playwright and physicist.

Zheng Zhenduo——Member of the Chinese Academy of Sciences (Academician), Vice Minister of Culture, Writer, Translator and Archaeologist.

Yuan Hanqing——a Chinese Academy of Sciences (academician), organic chemist.

Pei Wenzhong——a Member of the Chinese Academy of Sciences (academician), paleoanthropologist, discoverer of the first Peking Man skull.

Wang Shuzhuang——a Physicist.

This list of professionals allows us to understand the important role played by scientists in the establishment and development of the Natural History Museum. Not only they guided the various works of the museum with scientific thinking, but also enabled the latter to have leading figures in the four fields of zoology, botany, paleontology and anthropology since its establishment, and then cultivated more outstanding scholars batch after batch. Accomplished academic leaders have laid a solid foundation for the continuous development of the museum's scientific research work.

At the same time, they have also established a good working mechanism for the museum: on the one hand focus on scientific research, and the other hand, obtain first-hand information through field investigations and specimen collection.

In May 1958, the construction of the main building of the Museum of Natural History was completed, and the name of the museum was inscribed by Guo Moruo, then President of the Chinese Academy of Sciences. In January 1959, the Central Museum of Natural History was moved from the old sites of Chuanxin Hall, Wenhua Hall, Divine Kitchen, and three bungalows inside and outside the Donghua Gate of the Forbidden City to No. 126 Tianqiao South Street, Dongcheng District, got opened to the public and started to receive the audience. In January 1962, it was named the Beijing Museum of Natural History. In January 2023, according to the approval of the central editorial office, the editorial committee of the Beijing Municipal Party Committee officially approved and agreed to change the name of "

1958年5月，陈列楼落成。中国科学院院长郭沫若题写馆名"中央自然博物馆"

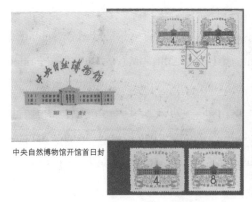

中央自然博物馆开馆首日封

《北京自然博物馆正式建馆》新闻报道

Chapter 1 The Discourse of Green Life

Beijing Museum of Natural History" to "National Museum of Natural History".

After decades of construction and development, the National Museum of Natural History has achieved fruitful results in its specimen collection, scientific research and popularization of paleontology, animals, plants, anthropology, and other earth sciences and life sciences. There are more than 370,000 pieces of collections in the existing collection, and the number of rare specimens ranks first in domestic natural history museums. In recent years, international academic journals of Nature and Science have published more than ten papers and won more than ten important awards such as the first prize of Beijing Science and Technology Award. The basic display continues to change, and temporary exhibitions and touring exhibitions with different themes are launched annually. Many popular science education activities such as "Museum Night" and "Small Science Popularization Explainer", which were first planned and launched in China, have formed characteristic brands.

As a bridge between human society and the natural world, the Museum of Natural History is a platform for dialogue between man and nature. It is playing an increasingly decisive role in coping with the challenges of the ecological crisis of the Earth and has begun to play an important role as an advocate and disseminator of protecting our planet.

The listed museums, based on the positioning and function of national museums, will further enrich and expand the collection of specimens, actively build research museums, carry out in-depth education planning and curriculum system design, strengthen academic exchanges at home and abroad, and comprehensively improve the overall capacity.

These exhibits have witnessed the vicissitudes of the Earth's history, and at the same time, they also silently interpret the arduous struggle of generations of scientists for the cause of Chinese museums. Each exhibit has its story that belongs to them as well as the scientists behind them.

Shortly, a new museum covering an area of 5.77 hectares, an area of about 200,000 square meters, will stand in the capital Beijing and will become a landmark building on the south-central axis.

The students are honored to be the little science guides of the National Museum of Natural History. For this reason, we are very proud!

… Chapter 1 The Discourse of Green Life

The Discourse of Green Life
(Exhibition Commentary)

The World of Plants

Luxuriant grass flourishes on meadowland.
Every year it blooms and then withers away.
Blazing wildfires are unable to wipe it out.
Under the spring wind's caress, it grows again.

The vibrant plants, intertwined in every corner of the earth, relate to hundreds of millions of creatures. Now, let's go into "The World of Plants" of the National Museum of Natural History and find the green life on this blue Earth.

The Evolution of Plants

Part one:

    Hello, everyone! Welcome to the "The World of Plants" exhibition hall of the National Museum of Natural History. I am XXX, a little science commentator, and I will explain everything to you today. "The World of Plants" is one of the four classic permanent exhibitions that the National Museum of Natural History has since its establishment. More than 1,200 plant fossils and various existing plant specimens are exhibited in the exhibition hall, including more than 30 kinds of national protected, endemic, and precious relic plants.

    "The World of Plants" is divided into three exhibition halls, namely "The Evolution of plants", "Diversification and Adaptation of Angiosperms", and "Plants and Humans". Together, they will present the green epic of the plant world to the audience. Next, please follow me into the exhibition hall and explore the mysteries of plants together.

生命之绿 *The green of life*

Part two:

  The Earth formed about 4.6 billion years ago. When it was just formed, it was full of gushing magma. A large amount of gas erupted from the magma. As the temperature dropped, it cooled down into rain and fell to the surface. It all gathered in low-lying places and gradually formed a primitive ocean.

  Cyanobacteria appeared around 3.5 billion years ago. They are the earliest known oxygen-producing photosynthetic microorganisms on the Earth and the only producers of free oxygen in its early atmosphere.

  Nostoc, also known as ground hair vegetable, hair algae, or asparagus, is a kind of algae of the order Nostoc of Cyanobacteria. It is widely distributed in deserts and barren soils all over the world. Because of its dark color, slender shape, and similarity to human hair, it is called "Facai" (rich) and is a first-class protected wild plant in my country. "Facai" has a certain nutritional value, and its name is homonymous with "Facai", so it is often served on human tables. In the past, nearly 200 million acres of grassland in Inner Mongolia were destroyed and decertified due to the large-scale collection and sale of wild nostocs. The cyanobacteria that have coexisted with drought for hundreds of millions of years are almost extinct.

  Red tides and algal blooms are a natural ecological phenomenon in which algae multiply in freshwater bodies, and they are one of the characteristics of water body eutrophication. After the wastewater containing nitrogen and phosphorus produced by domestic and industrial and agricultural production enters the water body, blue-green algae, green algae, diatoms, etc. multiply, making the water body hypoxic and appear blue or green, thereby causing deterioration of water quality and the death of aquatic animals.

"Stromatolites" are evidence of the life activities of cyanobacteria and bacteria. The substances produced in the life activities of cyanobacteria absorb calcium and magnesium ions. The periodic mineral deposition of calcium and magnesium ions makes the stromatolites present a layered structure and beautiful appearance after being polished. Patterns are sometimes used as building decoration materials.

There was no oxygen in the Earth's primitive atmosphere. The appearance of cyanobacteria changed the direction of biological evolution and accelerated the evolution of the Earth's surface environment and laid the foundation for the emergence of various types of organisms.

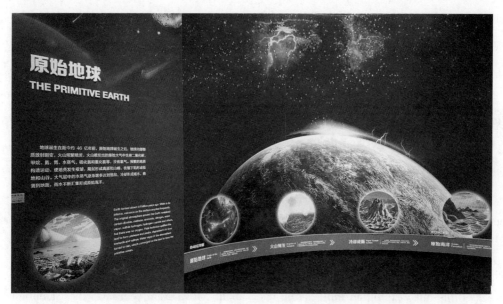

Part three:

Nowadays, there are about 30,000 species of algae on the Earth that are capable of photosynthesis. Unicellular eukaryotic red algae are considered the most primitive type of algae. They appeared about 1.9 billion years ago. By 1.6 billion years ago, primitive multicellular red and green algae had appeared. In the Cambrian period, cellular algae began to flourish, and the evolutionary trend of various main groups of algae was formed.

Everyone has eaten sea sedge, right? Most of the sea sedge is processed by *Porphyra yezoensis*. Laver is not a single species, but a general term for hundreds of species of algae in the Rhodophyta family Rubiaceae. We can say that sea sedge is a kind of laver, but not all laver can be processed into sea sedge. The laver in the soup we usually drink is called altar laver, which is usually dried and processed into discs for sale.

Day Lily is the main raw material of agar (it is somewhat similar to the jelly texture familiar to the audience), because agar has solidification and stability, and is widely used as a bacterial culture medium.

Sargassum kjellmanianum Yendo is a common species along the coasts of the Yellow Sea and Bohai Sea in China. In the past, it was only used as a fertilizer. Later, research proved that the content of alginate in *Sargassum kjellmanianum* Yendo is as high as 30%. At present, it has become one of the main raw materials to produce alginate. Alginate has a wide range of uses and can be implemented in the pharmaceutical industry, reactive dyes, water descaling, and other fields, and can be used as a familiar food thickener and emulsifier in food processing.

Part four:

Bryophytes appeared 480 million years ago and evolved from algae. They are the "hobbits" in the plant family. They have no vascular structure in their bodies and poor water transport capacity, so they are short. Bryophytes live in moraine deserts, tundra areas, exposed stone surfaces or fractured rock formations. They can secrete acidic substances to corrode the rock formation surface, so they are called the pioneers of the plant kingdom. Bryophytes are divided into three categories: mosses, liverworts, and hornworts.

On this huge exhibition wall, dating back about 540 million years ago the Cambrian "Underwater Forest" is hand-painted. In the primitive ocean, clusters of green algae, red algae, and brown algae grow on the seabed, forming a dense algae forest, which makes the vast expanse of blue ocean present a colorful scene.

This a *Sargassum kjellmanianum* Yendo up to four meters and seaweed up to two meters. These specimens bear a special historical significance as they pay homage to China's first systematic botanical exhibition. Samples were donated to us in 1959 by the Institute of Oceanic Studies of the Chinese Academy of Sciences in honor of the 10th anniversary of the People's Republic of China. They were returned to the exhibition hall after a half-century

Chapter 1 **The Discourse of Green Life**

of unrest. Although the algae have cracks, the traces of time have not weakened their charm, but on the contrary, added a heavy sense of the times to these ancient specimens.

Part five:

The Psilophytes exhibit showcases a terrestrial plant landscape from the Early Devonian period, some 400 million years ago. In the desolate swamp environment, there are dwarf naked ferns, such as Apex fern, artichoke fern, spiny fern, and so on. The appearance of these early land plants added new colors to the Earth and formed the early land vegetation. The process of plants transitioning from aquatic to terrestrial is an important link in the evolution of the Earth's terrestrial ecosystem. Since then, plants landed on land and begun a long and gorgeous life.

Lycopodiums are a special and diverse group of terrestrial vascular plants that arose about 400 million years ago. In the Carboniferous period 300 million years ago, it spread all over the world and is one of the main coal-making plants. Existing lycopodium plants only have four major categories: Lycopodium, Selaginaceae, Huperzaceae and Isoetaceae.

Plants vary in water content. The water content of aquatic plants is as high as 98%; the water content of woody plants is about 40% to 50%; the water content of herbaceous plants is about 70% to 80%; while the water content of plants in desert areas is only 6%.

The fairy grass that can bring the dead back to life in martial arts novels is described according to the characteristics of *Selaginella*. *Selaginella* is also called Nine Dead Resurrection Grass, and its vitality is very tenacious. Even if the water content drops below 5%, it can still maintain life. At this time, the leaves of *Selaginella* will curl up into fists due to drying. At first glance, they seem to have dried up, but once they encounter moisture, they can be stretched and revived.

Isoetes sinensis Plamer is an endangered aquatic fern unique to China. It belongs to the national first-level key protected wild plants and is known as the giant panda in the plant kingdom. Isoetes is the only surviving remnant genus in the family Isoetaceae which has important scientific research value. It is also a relic plant that survived Quaternary glaciers. It prefers a mild and humid climate. It grows in the silt in shallow ponds and ravines that are inaccessible. It is mainly distributed in some areas in the lower reaches of the Yangtze River.

Fems are widespread on the Earth, and it is an important coal-forming plant in the geological period. At present, there are more than 10,000 species of ferns on the Earth. Cuneiphytes are the most basic group of true ferns, and they are also ancient primitive vas-

cular plants.

Part six:

For most ferns, the new leaves are rolled into circles, and there are only a few more primitive types with less obvious curls. When a person falls, the legs curl up due to a natural reflex, just like the young leaves of a fern curl up before they open.

Here are the fossils of *Cladophlebis* Brongniart, which have been found in the Jurassic strata in the Fangshan area of Beijing. Friends who are interested in fossils can find them if have the opportunity.

The Sphenophytes mainly include *Calamites* Suckow and *Neocalamites*, which are characterized by having internodes on the stem, and the leaves are whorled between the nodes. This piece of wood, which looks a bit like bamboo at first glance, is a fossil of *Neocalamites*. Shen Kuo, a scientist in the Northern Song Dynasty, recorded this kind of plant in his book "Mengxi Bi Tan" as early as a thousand years ago, and conducted a study on the paleoenvironment based on fossils. It is speculated that this is very instructive for the study of modern environmental science.

The Carboniferous period, about 350~290 million years ago, is a representative period of plant flourishing. At that time, the Earth's climate was warm and humid, with swamps

everywhere, and various plants were extremely dense. After the plants in the forest die, they are buried in the swamp for a long time, isolated from the air, and carbonized and metamorphosed into coal. Therefore, the Carboniferous period is also the most important coal-forming period in geological history.

Part seven:

Alsophila Spinulosa, also known as "Tree Fern", is the only woody fern that can grow into towering trees on the earth today. They appeared at the same time as dinosaurs, and the *Alsophila Spinulosa* was one of the main foods of herbivorous dinosaurs. Over time, the environment on the Earth has changed, and natural events such as crustal movement and glacier eruptions have occurred one after another. Many animals and plants have suffered extinction, but *Alsophila Spinulosa* survived and became a very small number of relic plants and got the reputation of a "Living Fossil". Nowadays, the *Alsophila Spinulosa* can barely survive in the forests of the tropics, and most of its members are facing extinction. The deforestation of tropical virgin forests by humans and the use of *Alsophila Spinulosa* to make handicrafts are all accelerating the demise of this group.

Pregymnosperms were a group of special terrestrial vascular plants that once existed on Earth. They appeared more than 400 million years ago and disappeared from the earth about 300 million years ago. Pregymnosperms have both true fern and gymnosperm characteristics. Its existence shows that gymnosperms are likely to have originated from true ferns.

Cycadophytes first appeared in the Carboniferous period, flourished in the Jurassic to Cretaceous periods of the Mesozoic Era, and then began to decline. Some people call the Mesozoic Era "the Age of Cycads". There are two families, ten genera, and more than 300 species of living cycads, which are distributed in tropical and subtropical regions. In my country, there are one family, one genus, and thirty species, which are distributed in Southwest and South China.

In the middle of the showcase, this Cycad fossil with a diameter of more than one meter comes from the Late Triassic strata in the Beipiao area of Liaoning of my country. It was discovered in 2010 by Mr. Li Chengsen, the former curator of our museum and doctoral supervisor, and named it "Leptocycas *yang caogouensis* Zhang et al." The fossils are beautifully preserved, with stretched pinnate leaves branching spirally and radiating from a central stem in which the reproductive structures are preserved. In the showcase, the contrastive display of living Cycads and extinct Cycad fossils allows the audience to see the evolution process of their origin, prosperity, and decline.

Ginkgo Biloba belongs to gymnosperms, a unique plant type in my country, and is also a common green street tree. It first appeared in the early Permian period, about 290 million years ago. There are many kinds of them, and they used to spread all over the world. After the Cenozoic Era, most species became extinct. Now there is only one family of Ginkgoaceae and one species of *Ginkgo Biloba*.

Conifers are the most diverse and widely distributed category of gymnosperms. The living conifers include seven families and more than 600 species. The metasequoia and cathaya exhibited here are national first-class protected plants, and the Picea meyeri, Pseudolarix amabilis, Glyptostrobus pensilis, Pinus bungeana and Pinus kwangtungensisare all unique plants in China.

Most conifers are green all year round. So, let me ask a question, will the branches and leaves of evergreen plants age? The branches and leaves of coniferous trees such as pine, fir, and cypress will age, and the aged branches and leaves will fall off. The branches and leaves of coniferous trees gradually age, and the old leaves gradually fall off. Compared with the huge green crown, this slow aging and falling off phenomenon is easy to be ignored, so it gives people a visual impression of evergreen all year round.

Rattans, also known as lid plants, are a special type of gymnosperm, named for their

special organ, the lid. There are more than 100 kinds of vine plants in existence, including the three groups of vines, ephedra, and orchid. Over the past century, hemp vines have played a pivotal role in the study of the origin of angiosperms. In clinical medicine, ephedra is often used to treat colds, coughs, and asthma. It is rich in alkaloids and is the main resource for extracting ephedrine. For example, the main ingredient of Contac that we are familiar with is pseudoephedrine hydrochloride.

Part eight:

Angiosperms are the highest group of plants. They are called flowering plants because of their unique reproductive organs–flowers. Angiosperms may have originated in the Jurassic period or earlier but did not flourish on Earth until about 80 million years ago. After entering the Cenozoic Era, angiosperms continued to rapidly differentiate and eventually became the absolute protagonists in the plant kingdom. At present, there are more than 200,000 species of angiosperms in the world, which is the largest category of living plants. So, what did the earliest flowers in the world look like, and in which part of the Earth did they first bloom?

Chapter 1 The Discourse of Green Life

This question haunted Darwin, a famous British biologist, more than 100 years ago. He found that about 100 million years before us in prehistoric times, flowers had spread to every part of the earth. However, going back further, these flowering plants mysteriously disappeared, leaving no evidence of their evolution at all. How did flowering plants appear and how did they evolve? Darwin dubbed it a "nasty riddle" after extensive investigation and no clues. From 1998 to 2002, Sun Ge, a Chinese paleobotanist, led a research team to discover for the first time the earliest angiosperms in the world at that time——*Archaefructus liaoningensis* and the "*Archaefructus Sinensis*" fossil found fossil evidence for the early evolution of flowers.

Next, please follow me to the exhibition hall on the left to see "Plants and Humans Beings".

Plants and Human Beings

Part one:

Welcome to the "Plants and Human Beings" exhibition hall. The relationship between plants and humans is inseparable. From picking fruits to making fire by drilling wood, from textiles to construction, human beings cannot live without plants.

Here, we can see that the original wild *Oryza sativa. L* produced few seeds, while the bred *Oryza sativa. L* produced more seeds. Now, let me ask you, which part of the *Oryza sativa. L* is the rice we eat? (Endosperm of the seed)

Do you know what the "Five grains" we often eat refer to? The five grains mentioned by the ancients are not the same as today's five grains. Today's whole grains mostly refer to rice, wheat, sorghum, soybeans, and corn.

In ancient times, there was a legend that Shennong tasted a hundred herbs to cure diseases. Today, Chinese scientist Tu Youyou's team has extracted artemisinin from plants and saved the lives of millions of malaria patients. In plants, there are many treasures worthy of

Chapter 1 The Discourse of Green Life

human excavation.

Traditional Chinese medicine culture is an important part of Chinese traditional medical culture, and its origin can be traced back thousands of years. In primitive societies, people used wild herbs to treat diseases. During the Shang and Zhou dynasties, traditional Chinese medicine began to form a system. Here are some well-known Chinese patent medicines, such as *Glycyrrhiza uralensis* Fisch, *Rehmannia qlutinosa*, *Eriobotrya japonica*, etc.

Part two:

Oil plants can not only provide edible oils such as peanut oil, corn oil, and soybean oil, but also provide industrial oils, such as castor oil and tung oil. Tung oil is an important industrial oil plant, and its seeds are squeezed out to be excellent dry vegetable oil. Tung oil is widely used and is the main raw material for the manufacture of paints and inks. Its oil yield is high, five-six times that of peanuts and ten times that of soybeans, and it is known as the "king of oil in the world".(please follow me to the center of the exhibition hall.)

Sabal palmetto Lodd. belongs to the palmae and is an evergreen tree native to tropical regions of Asia. The height of the tree is about twenty meters, the trunk is straight and round, without branches, the crown is like a giant umbrella, and the leaves are spread out like palms. It is an excellent tree species for greening environments in tropical areas.

The green of life

The Patta-leaf scriptures written on the leaves of the tree, with a history of more than 2,500 years. It originated from ancient India, has a very high cultural value, and has the reputation of "Buddhist Panda". Patta-leaf scriptures is written on the leaves with iron hairpins and then painted with paint. The writing is clear and cannot be erased. After special treatment such as boiling, it can also prevent insects, water, and deformation. This is why the Patta-leaf scriptures can be preserved for thousands of years without decay. China's Tibet is the region with the most preserved Patta-leaf scriptures in the world today. At the same time, the Patta-leaf scriptures is also an important source material for the study of ancient Tibetan religion and art.

Part three:

Here is a specimen of the fiber plant. There are hundreds of species of fiber plants all over the world. Although synthetic fibers are now widely used in our lives, industries such as papermaking still cannot do without fiber plants. Before the invention of synthetic fibers, people's clothes were inseparable from plant fibers, and the closest thing to us humans is cotton. Shown here is a set of men's bark clothes, which are made by ethnic minorities in the southwest of our country from the bark of the *Antiaris toxicaria* Lesch.

The milky white sap of the *Antiaris toxicaria* Lesch is highly poisonous. Once it comes into contact with wounds of humans and animals, it paralyzes the heart of the poisoned person, coagulates their blood, and even suffocates them to death. Because the bark of the *Antiaris toxicaria* Lesch is poisonous, it has the characteristics of insect resistance and corrosion resistance. It is used to make clothes.In addition, the fiber is soft, easy to wear, and washable, so it is loved by the ancestors. Although nature has created such highly poisonous plants, as long as they are used properly, they can also benefit human beings.

Part four:

  Spice plants are a class of plants that can produce and extract aromatic substances from them. There are more than 600 types of aromatic plants in my country.

  When it comes to spice plants, people generally think of perfume. In addition to perfume, there are plant essential oils, peppermint oil, cooling oil, camphor, etc. The most closely related to us humans are edible spice plants, such as pepper, aniseed, star anise, cassia bark mustard, etc., which appear on people's tables almost every day.

  There are many secondary metabolites in the roots, stems, leaves, flowers or fruits of plants. Some of them can color other materials, so they are called dye plants. Natural pigments such as chlorophyll, carotene, and tomato red can be processed into food pigments to make food look more delicious and attractive. At the same time, plant dyes are also widely used in works of art such as traditional Chinese painting, thangka, porcelain, and batik.

生命之绿 *The green of life*

  Now, everyone please follow me to the exhibition hall on the right to continue to study plants, look at the flowers and plants around us, and learn about "dirersification and adaptation of angiosperms".

第 1 章 绿色生命的倾诉
Chapter 1 The Discourse of Green Life

Diversification and Adaptation of Angiosperms

Part one:

Welcome to the "Dirersification and Adaptation of Angiosperms" exhibition hall. Angiosperms are also called flowering plants, and most of their seeds are enclosed in fruits. It is a group with the highest level of evolution, the largest variety and the widest distribution in the plant kingdom. There are more than 200,000 species of angiosperms in existence, accounting for about half of the plant kingdom, and there are more than 30,000 species in my country.

Why are angiosperms so prosperous, and what kind of structure and survival skills do they rely on to spread to every corner of the earth, adding colorful colors to the world? Now, let us walk into the world of angiosperms together.

Part two:

Angiosperms have six major organs: roots, stems, leaves, flowers, fruits and seeds. The first thing I want to introduce to you is the leaves of angiosperms. Leaves carry out photosynthesis through chloroplasts to produce organic matter, and leaves are responsible for transpiration and respiration. It is an important vegetative organ of plants. A few plants can also reproduce through leaves. For example, many of the succulents we are familiar with are rooted (cuttings).

Generally, the leaves of plants can be divided into three parts: blade, petiole and stipule. If there are three parts, it is called a "complete leaf", such as the leaves of plants such as roses; if it lacks petioles or stipules, it is called "incomplete leaves", such as plants such as cabbage. Different plants have different leaf shapes.

On this exhibition wall, a total of 137 wax leaf specimens of 127 species of angiosperms in different shapes are exhibited. Their leaf shapes, leaf margins, and leaf tips are all different, which is the best evidence that plants have undergone diverse evolution to adapt to different environments. Seeing so many leaves, have you ever wondered why some

leaves are green and others are red? The leaves of most plants contain the most chlorophyll, so they will appear green. When autumn comes, the leaves of *Ginkgo biloba* L. and *Acer buergcrianum* Mia. will become yellow; the leaves of *Liquidambar formosana Hance* and *Cotinus coggygria* Scop. will turn red. Because their leaves contain not only chlorophyll, but also carotenoids, phycoerythrin or anthocyanins.

Does the similar leaf shape mean that the plants are related? Not necessarily. The roots, stems, and leaves of plants change their shape with the environment. Different plants grow in the same environment, and after a long period of evolution, they may also evolve similar leaf shapes. Relatively speaking, the same plant growing in different environments may evolve leaves of different shapes, thicknesses and sizes. Therefore, to identify whether plants are related, it must be judged from the aspects of flower shape, stamens, trees and arrangement.

After understanding the leaves of angiosperms, let's take a look at the flowers of angiosperms. Generally, flowers consist of corolla, calyx, receptacle, and pistil. Flowers are an important reproductive organ of angiosperms, with bright colors and a wide variety of flowers adorning the beautiful earth.

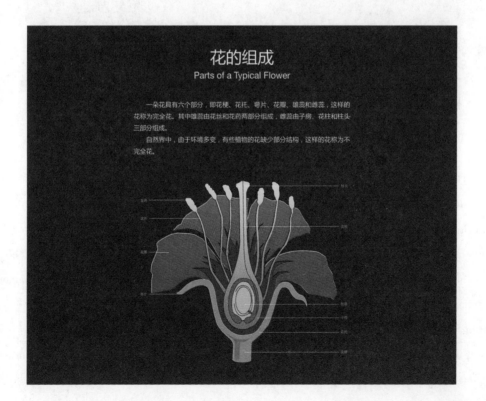

The green of life

In this unit, 18 preserved leaf specimens and 28 soaked specimens of flowers are displayed. The so-called inflorescence is the sequence of different forms of flowers on the flower axis. This wall introduces the secrets of flower pollination to everyone in the form of pictures and multimedia videos.

Here is a dipped specimen of the fruit of the plant. Ordinary plant specimens are faded and shriveled, but now "chemical replacement color preservation technology" is used to make soaked specimens, and chemical molecules are used to replace and fix various colored substances in plants, which can keep the color of plants for ten to fifty years.

Part three:

There are 181 species of fruits and seeds on display, the most intriguing of which are the Infructoses of the *Raphia vinifera* Beauv. which belongs to the palmae. Its shape is thick and drooping, like an elephant's nose, and its fruit is oval or obovoid. Botanically, *Raphia vinifera* Beauv is called a flowering and fruiting plant, only blooms once in its life, bears fruit once, and then the whole plant dies. Its lifespan is about twenty years, and among perennials, once-flowering plants are rare. *Raphia vinifera* Beauv is native to the tropical regions of Africa. It has been introduced and cultivated in Xishuangbanna in Yunnan, Nanning in Guangxi and Taiwan. This precious giant specimen was collected from the Xishuangbanna Botanical Garden of the Chinese Academy of Sciences.

Chapter 1 The Discourse of Green Life

Plate roots and strangulation are two unique phenomena of tropical rainforests. The plank roots are developed from the thick lateral roots close to the ground. The trees in the tropical rainforest are tall. The existence of the plank roots increases the supporting capacity of the trees, which can solve the problem of "top-heavy" and unstable standing of tall trees.

Strangulation is a common phenomenon in tropical rainforests. For example, the seeds of the plants of the genus Moraceae will be spread to other trees by birds and other small animals. After the seeds germinate and grow up, they will deprive the host plant of water and nutrients. The scarcity, death, and rot of the plants have become the new masters of the "indomitable" here. To compete for the precious sunlight and limited living space in the rainforest, plants are engaged in a silent and cruel struggle every day.

Parashorea chinensis Wang Hsie, is a rare tree species unique to Yunnan, China, with a height of more than forty to sixty meters. They are only distributed within the rainforest of twenty square kilometers in Xishuangbanna. This species has been listed as a national first-level key protection wild plant.

Part four:

To adapt to the harsh living environment of the desert, most desert plants have low plants and many branches in the lower half. The developed root system helps absorb soil moisture, and the small and narrow leaves can reduce water evaporation. These characteristics can help desert plants effectively resist desert areas' drought and sandstorms.

This is the stem of *Populus euphratica* donated by Mr. Battelle of Alxa League in Inner Mongolia. *Populus euphratica* is the only natural tree in the desert. It is known as the " Backbone of the Desert" and "the Hero Tree of the Desert". It can resist drought, wind and sand, and salt and alkali.

Populus euphratica has a beautiful legend of 3,000 years. It is said that it can live without dying for a thousand years, die without falling for a thousand years, and be immortal for a thousand years. From this, we can feel the tenacious vitality of *Populus euphratica*. In the specimens, you can see that the leaf shape of the same tree may be different in different growth periods and different growth parts. Why are the leaf shapes of *Populus euphratica* so different? This is related to its growth environment. Desert areas are extremely short of water, and the young leaves are small, long and narrow like willow leaves, which can reduce the rate of photosynthesis, reduce transpiration, and save water. When the roots have penetrated deeply into the ground and the tree has absorbed enough water, the leaves will gradually become larger and become broadly oval or heart-shaped, as round as poplar leaves.

第 1 章　绿色生命的倾诉
Chapter 1　The Discourse of Green Life

Part five:

Here, I will introduce Mangroves to you. Mangroves are not red trees. Because these mangrove plants are mainly distributed on tropical and subtropical beaches, they are collectively called "Mangroves". When typhoons and tsunamis come, Mangroves are a natural barrier, protecting the earth from wind and waves. Because of their existence, many coastal cities have escaped natural disasters, so they are called "wave–dissipating pioneers" and "coast guards".

The most wonderful way of reproduction of Mangroves is "viviparous". Viviparity is a special reproductive method of mammals. Simply put, it means that the mother directly gives birth to the baby from the womb. How did the mangrove plants "viviparous"? It turns out that because the mangroves have been subjected to the tide and waves for a long time, it is difficult for the seeds to have a stable germination environment. Therefore, the seeds of many plants in the mangroves have germinated and grown into small ones inside the fruit before leaving the mother. Seedlings continuously absorb nutrients from the seeds, and slowly turn into small awls like "pens". When the time is right, it will fall straight down and insert itself into the silt of the beach, and it will soon take root and grow into a new plant.

红树林植物
Mangrove Plants

生命之绿 *The green of life*

Part six:

The alpine region has low temperatures all year round, the living environment is extremely harsh, and the temperature will drop by about 0.6 degrees for every hundred meters above sea level, so alpine plants generally grow close to the ground. *Thylacospermum caespitosum* Camb is a cushion plant that can reach a diameter of more than 1.3 meters, with branches tightly packed together to prevent heat and water loss and adapt to cold and windy weather. Therefore, they are also known as the "king of cushion plants".

Now, let's take a look at the "smart" group—insectivorous plants, which are a class of plants that capture and digest animals and obtain nutrition from them. They belong to a special group in the plant kingdom, and there are about 500 species in the world. Insectivorous plants can obtain nutrients through photosynthesis. Most of them grow in poor soil, especially in areas lacking nitrogen, and need to catch insects to supplement the nutrients needed for plant growth and development.

In summer, people may buy a few pots of *Nepenthes mirabilis* Lour, hoping that they can help us "eat some mosquitoes". This trick is not effective, because the female mosquitoes that rely on blood-sucking for survival are not interested in honey liquid and ultraviolet rays, so they will not fly to the pitcher plant to throw themselves into the net. Let me test everyone, what is the function of the lid of the *Nepenthes mirabilis* Lour bag? The function of the cover is only for decoration, and it will not be covered at all (at most, it will block the rainwater to prevent the digestive juice from being diluted). Insects are attracted by the honey juice at the mouth edge of the insect trap, while sucking the honey juice, they walk to the slippery interior of the insect trap and finally throw themselves into the net.

Next, Let us introduce a relatively selfish plant—parasitic plants, which feed on living organisms and obtain all or most of the nutrients and water they need from green plants. For example, *Boschniakia rossica* cannot carry out photosynthesis, they parasitize the roots of alder plants. Traditional Chinese medicine believes that *Boschniakia rossica* can strengthen the body and nourish the body. In the past, it was often stolen. *Boschniakia rossica* belongs to the second-class protected plants in the country. Human beings should not destroy the ecological environment of the earth for personal desires.

After listening to my explanation, do you have a certain understanding of plants? Thank you, that's all for my explanation!

生命之绿 The green of life

走进国家自然博物馆

国家自然博物馆是中国唯一的国家级、综合性自然博物馆，代表国家保护、研究、收藏、阐释、展示自然和人类发展过程中具有历史、科学和艺术价值的自然遗产。

亲爱的同学们，每当我们漫步在国家自然博物馆的展厅中，驻足琳琅满目的展品，感叹生物进化的神奇时，我们是否想过，在了解地球历史变迁的同时，也了解一下国家自然博物馆的发展历史呢？

现在，我们就来到首都北京南城的中轴线上，看看这座中华人民共和国依靠自己的力量建立的第一座大型自然历史博物馆。

它的前身是成立于1951年4月2日的中央自然博物馆筹备处。当时，筹备处有7位科学家和5位文化部领导。

7位科学家如下：

胡先骕　中央研究院（院士），中国植物分类学的奠基人。

郑作新　中国科学院学部委员（院士），中国现代鸟类学的奠基人、动物地理学的开拓者。

李继侗　中国科学院学部委员（院士），中国植物生理学的开拓者、植物生态学与植物学的奠基人之一。

高振西　中国科学院学部委员（院士），中国地质博物馆和中国第一个国家级地质自然保护区的主要创建人。

孙云铸　中国科学院学部委员（院士），中国古生物学奠基人、中国地质学会的创始人之一。

李　璞　教授，著名地质学家、中国同位素地球化学的开创者。

张春霖　教授，中国现代鱼类学的奠基人。

5位文化部领导也都是学贯中西、文理兼修的大家：

丁西林　剧作家、物理学家。

郑振铎　中国科学院学部委员（院士）、文化部副部长、作家、翻译家、考古学家。

袁翰青　中国科学院（院士），有机化学家。

裴文中　中国科学院学部委员（院士），古人类学家、第一个北京猿人头盖骨的发现者。

王书庄　物理学家。

第 1 章　绿色生命的倾诉
Chapter 1　The Discourse of Green Life

　　这份大家云集的名单,让我们了解到科学家在自然博物馆的创立和发展中发挥的重要作用。他们不仅用科学的思维指导博物馆的各项工作,也使博物馆从创立伊始,就在动物学、植物学、古生物学和人类学四大领域拥有了领军人物,继而培养出一批又一批颇有成就的学科带头人,为博物馆科研工作的不断发展奠定了坚实的基础。

　　同时,他们也为博物馆建立了良好的工作机制:一手抓科研,通过野外考察、标本采集获得第一手资料;一手抓科普,将科研成果转化为陈列、展览、图书、讲座等。

　　1958 年 5 月,自然博物馆的主体建筑落成,由时任中国科学院院长郭沫若题写馆名。1959 年 1 月,中央自然博物馆由故宫东华门内的传心殿、文华殿、神厨和三座门内外的平房等旧址迁移到位于东城区的天桥南大街 126 号,并开馆对外接待观众。1962 年 1 月,定名为北京自然博物馆。2023 年 1 月,按照中央编办批复,北京市委编委正式批准同意"北京自然博物馆"更名为"国家自然博物馆"。

　　经过几十年的建设发展,国家自然博物馆在古生物、动物、植物和人类学等地球科学、生命科学领域的标本收藏、科学研究和科学普及工作取得了丰硕成果。现有馆藏藏品 37 万余件,珍稀标本数量在国内自然博物馆居首位。近年来在国际学术期刊 Nature 和 Science 累计发表论文十余篇,荣获北京市科学技术奖一等奖等重要奖项十余项。基本陈列持续改陈,每年推出不同主题的临时展览和巡回展览。在国内率先策划推出的"博物馆之夜""小小科普讲解员"等众多科普教育活动已经形成特色品牌。

　　自然博物馆作为连接人类社会与自然世界的桥梁,是人与自然"对话"的平台,在应对地球生态危机挑战中,愈来愈发挥着举足轻重的作用,并开始担纲起守护地球的倡导者和传播者的重要角色。

　　这座入选"中国首批 20 世纪建筑遗产"名录的博物馆,将立足国家馆的定位与职能,进一步丰富扩充馆藏标本,积极建设研究型博物馆,深入开展教育规划和课程体系设计,加强国内外学术交流,全方位提升整体能力。

　　这些见证了地球历史沧桑的展品,同时,也默默无声地诠释着一代代科学家们为了中国博物馆事业呕心沥血的艰苦奋斗历程。每一件展品都有一段属于它们,以及它们背后的科学家们的故事。

　　在不久的将来,一座占地 5.77 公顷、面积约 20 万平方米的新馆将矗立在首都北京,成为未来南中轴线上的标志性建筑。

　　同学们有幸成为国家自然博物馆的小小科普讲解员,为此,我们深感骄傲与自豪!

生命之绿
The green of life

绿色生命的倾诉
（展厅讲解词）

离离原上草，一岁一枯荣。野火烧不尽，春风吹又生。生机勃勃的植物，在地球的每一个角落中盘结，与亿万生灵息息相通。现在，就让我们走进国家自然博物馆的植物世界，找寻这颗蔚蓝色地球上的绿色生命……

植物演化

○ 第一段

各位观众大家好，欢迎参观国家自然博物馆"植物世界"展厅，我是小小科普讲解员XXX，今天由我来为大家讲解。"植物世界"是国家自然博物馆自建馆以来一直保留的四大经典常设展陈之一。展厅里展出了 1200 多件植物化石和各类现存植物标本，其中国家重点保护、特有及珍贵孑遗植物达 30 多种。

"植物世界"共分为 3 个展厅，分别是"植物演化"、"被子植物的繁盛与适应"、"植物与人类"，它们共同为观众朋友们展示植物世界这部绿色的史诗。接下来，就请大家随我走进展厅，一起探索植物的奥秘。

○ 第二段

地球诞生于约 46 亿年前，在它刚刚形成的时候，到处都是喷涌的岩浆。从岩浆中喷发出大量气体，随着温度降低，冷却成雨水并降落到地表，聚集在低洼处就渐渐形成了原始海洋。

蓝细菌大约出现在约 35 亿年前，它是地球上已知最早的产氧光合微生物，也是早期地球大气自由氧的唯一生产者。

发菜又叫地毛菜、头发藻、龙须菜，是蓝菌门念珠藻目的一种藻类。它广泛分布于世界各地的沙漠和贫瘠土壤中，因颜色黝黑，形状细长，和人的头发相似，所以被称为"发菜"，是我国一级重点保护野生植物。发菜有一定营养价值，名字又与"发财"谐音，经常被端上人类的餐桌。过去，由于大规模采集和销售野生发菜，内蒙古近两亿亩草场遭到破坏和沙化。发菜这种与干旱共存了亿万年的蓝藻几乎绝迹。

赤潮、水华是淡水水体中藻类大量繁殖的一种自然生态现象，是水体富营养化的特征之一。生活及工农业生产产生的含有氮、磷的废污水进入水体后,蓝藻、绿藻、硅藻等就会大量繁殖,使水体缺氧,呈现蓝色或绿色,从而引起水质恶化和水生动物的死亡。

叠层石是蓝藻和细菌生命活动的证据,蓝藻生命活动中产生的物质吸附了钙镁离子,钙镁离子的周期性矿物质沉积,使得叠层石被抛光后呈现出层层构造和美丽的图案,有时也被用作建筑装饰材料。

地球的原始大气中是没有氧气的,蓝藻的出现改变了生物进化的方向,加速了地球表层环境的演化,为后续各类生物的出现奠定了基础。

○ 第三段

地球上已知的藻类植物约有 3 万多种,它们是能够进行光合作用的低等植物。单细胞真核红藻类被认为是最原始的藻类植物,出现在 19 亿年前;到了约 16 亿年前,原始多细胞红藻和绿藻出现;在 5 亿年前的寒武纪时期,海洋中的藻类已经非常繁盛,各大类群藻类植物的进化趋势基本形成。

大家都吃过海苔吧!绝大部分海苔都是加工过的条斑紫菜。其实,紫菜并不是一个单独的物种,而是红藻门红毛菜科中上百种藻类植物的统称,我们可以说海苔是一种紫菜,但并非所有的紫菜都能被加工成海苔。平时我们喝的紫菜汤里的紫菜叫作坛紫菜,一般是晒干加工成圆盘状售卖的。

黄花菜是琼脂的主要原料(就是和观众们熟悉的果冻质地有点类似),因为琼脂具有凝固性和稳定性,被大量用作细菌培养基。

海黍子是中国黄海、渤海沿岸较常见的种类。过去它仅仅作为肥料使用,后来研究证明海黍子中褐藻胶的含量高达 30%,目前它已成为制造褐藻胶的主要原料之一。褐藻胶的用途广泛,可用于医药工业、活性染料、水质除垢等领域,在食品加工中可以作为我们熟悉的食品增稠剂和乳化剂等。

○ 第四段

苔藓植物出现在 4.8 亿年前,由藻类植物演化而来,它是植物家族里的"霍比特人",体内没有维管束结构,输水能力较差,所以个体矮小。苔藓植物生活在沙碛荒漠、冻原地带及裸露的石头表面或是断裂的岩层上,它能分泌酸性物质腐蚀岩层表面,因此被称为植物界的拓荒者。苔藓植物分为藓类、苔类、角苔类三大类。

在这面巨型展墙上,手工绘制着约 5.4 亿年前的寒武纪"海底森林"。原始海洋中,丛丛簇簇的绿藻、红藻和褐藻生长在海底,形成了茂密的藻类森林,使碧波万顷的海洋呈现出一派五光十色的景象。

生命之绿 *The green of life*

这里是长达 4 米的海黍子和长达 2 米的海带。这些标本具有特殊的历史意义,它们在向中国的第一个系统植物学展览致敬。这些标本是 1959 年中国科学院海洋研究所在献礼国庆十周年时无偿捐赠给我馆的,经历了半个世纪的风雨,标本重回展厅。虽然藻体有了裂痕,但岁月的痕迹非但没有削弱它们的魅力,反而为这些古老的标本增添了厚重的时代感。

○ 第五段

裸蕨类植物展区展示了约 4 亿年前泥盆纪早期的陆地植物景观。在荒凉的沼泽环境中,生活着矮小的裸蕨类植物,如顶囊蕨、工蕨、刺镰蕨等。这些早期陆地植物的出现为大地增添了新的色彩,形成了地球早期的陆地植被。植物从水生过渡到陆生的这一过程,是地球陆地生态系统演化的重要环节。从此以后,植物登上了陆地,开始了漫长而绚丽的生活。

石松类植物是陆生维管植物中的一个特殊类群,它们种类多样,出现于约 4 亿年前。在约 3 亿年前的石炭纪就已经遍及世界,是主要的造煤植物之一。现存的石松类植物仅有石松科、卷柏科、石杉科和水韭科四大类。

植物的含水量各不相同。水生植物含水量高达 98%;木本植物的含水量约 40%~50%;草本植物的含水量约 70%~80%;而沙漠地区植物的含水量只有 6%。

武侠小说中能够起死回生的仙草,是根据卷柏的特征来描写的。卷柏又叫九死还魂草,它的生命力十分顽强。即使含水量降到 5% 以下,仍然可以保持生命。这时,卷柏的叶子会因干燥而卷成拳状,乍一看似乎已经干死,可是一旦遇到水分,就可以舒展复活。

中华水韭是中国特有的濒危水生蕨类植物,属于国家一级重点保护野生植物,被称为植物界的大熊猫。水韭属是水韭科中唯一生存的残遗属,有重要的科研价值。它也是经过第四纪冰川后残存下来的孑遗植物,喜欢温和湿润的气候,生长在人迹罕至的浅水池沼和山沟里的淤泥中,主要分布于长江下游局部地区。

真蕨类植物是地球上分布广泛的孢子植物,是地质时期重要的造煤植物。目前,地球上真蕨类植物有 1 万多种。楔叶类植物是真蕨中最基础的类群,也是古老的原始维管植物。

○ 第六段

大多数蕨类植物,新叶都是卷成一圈圈的,只有极少数较原始的种类卷曲不明显。人跌倒时腿会因为自然反射而卷曲,这就好比蕨类植物的幼叶,在没有张开之前都是卷曲的。于是,古人将"蹶"的足字旁去掉,加上草字头,"蕨"字就诞生了。

这边是枝脉蕨化石,在北京房山地区的侏罗纪地层中就发现过枝脉蕨的化石,对化石感兴趣的朋友们,有机会可以去找一找。

楔叶植物主要有芦木和新芦木,它们的特点是茎上具有节间,叶子在节间轮生。这块猛一看有点像竹子的是新芦木化石,北宋科学家沈括早在一千年前,就在他的著作《梦溪笔谈》中记载了这类植物,并根据化石对古环境进行了推测,这对现代环境学的研究很有启发意义。

距今约 3.5 亿年至 2.9 亿年的石炭纪,是植物大繁盛的代表时期。当时地球的气候温暖湿润,沼泽遍布,各类植物异常茂密。森林中的植物死亡后,长期埋藏在沼泽中,与空气隔绝,炭化变质成煤。因此,石炭纪也是地质历史时期最重要的成煤期。

○ 第七段

桫椤又称"树蕨",是现今地球上唯一能长成参天大树的木本蕨类植物。它们和恐龙同时出现,饱含淀粉的桫椤是植食性恐龙的主要食物之一。随着时间的推移,地球上的环境发生改变,地壳运动、冰川爆发等自然事件接连发生,许多动植物遭受了灭顶之灾,桫椤却死里逃生,成了数量极少的孑遗植物,有了"活化石"的美称。现在,幸存于地球上的桫椤只能在热带地区的森林中勉强生存,多数成员面临濒危灭绝的境地,人类对热带原始森林的砍伐,以及利用桫椤制造工艺品的行为,都在加速这一类群的灭亡。

前裸子植物是地球上曾经存在的一类陆生维管植物。它们出现于 4 亿多年前,在约 3 亿年前从地球上消失。前裸子植物既具有真蕨类植物的特征,又具有裸子植物的特征。它的存在,说明了裸子植物很可能起源于真蕨类植物。

苏铁类植物最早出现于石炭纪,繁盛于中生代的侏罗纪至白垩纪,之后便开始衰落,有人把中生代称为"苏铁植物的时代"。现生苏铁类植物有 2 个科、10 个属、300 多个种,分布于热带和亚热带地区,我国有 1 科、1 属、30 种,分布于西南和华南地区。

在展柜中间,这件直径 1 米多的苏铁植物化石来自我国辽宁北票地区晚三叠世地层,2010 年由我馆原馆长、博士生导师李承森先生发现,命名为"羊草沟纤苏铁"。化石保存得非常完美,舒展的羽状叶片从中间的茎螺旋状辐射分枝而出,茎中间保存了生殖结构。展柜中,现生苏铁类和灭绝苏铁化石的对比展出,可以让观众们清晰地看出苏铁类植物的起源、繁盛和衰落的演化过程。

银杏属于裸子植物,是我国特有的植物类型,也是常见的绿化行道树,最早出现于二叠纪早期,距今约 2.9 亿年,它们种类繁多,曾经遍布世界各地。新生代之后,大部分种类都灭绝了。现在仅存银杏科 1 科、银杏 1 种。

松柏类植物是裸子植物中种类最多、分布最广的一类。现生的松柏类植物包括 7 个

生命之绿 The green of life

科,有 600 多种。这里展出的水杉和银杉是国家一级保护植物,白扦、金钱松、水松、白皮松和华南五针松都是中国特有的植物。

绝大多数的针叶树都具有常绿性,它们终年翠绿。那么,我来问大家一个问题,常绿植物的枝叶会老化吗?其实,松、杉、柏等针叶树的枝叶都会老化,老化的枝叶会脱落。针叶树的枝叶是逐渐老化的,老叶逐步脱落,这种慢条斯理的老化和脱落现象,跟庞大的绿色树冠比起来,很容易被忽略,所以给人一种终年常绿的视觉印象。

买麻藤类植物又称盖子植物,它们是一类特殊的裸子植物,因其具有特殊的器官——盖子而得名。现存的买麻藤类植物有 100 多种,包括买麻藤、麻黄和百岁兰三大类群。过去一个世纪以来,买麻藤类植物在被子植物的起源研究中具有举足轻重的地位。在临床医学中,麻黄常常被用于治疗伤风感冒、咳嗽气喘,它的生物碱含量丰富,是提取麻黄碱的主要资源,比如,大家熟悉的康泰克的主要成份就是盐酸伪麻黄碱。

○ 第八段

被子植物是植物界中最高级的类群。它们因为具有独特的生殖器官——花,所以被称为有花植物。被子植物可能起源于侏罗纪甚至更早,但直到约 8 千万年前才在地球上逐渐繁盛起来。进入新生代之后,被子植物继续快速分化,最终成为植物界中的绝对主角。目前,世界上的被子植物有 20 多万种,是现生植物中数量最多的一类。那么,世界上最早的花是什么样子的,它们又绽放在地球的哪一个角落呢?

这个问题早在 100 多年前就萦绕在英国著名生物学家达尔文的心头。他发现,距离我们 1 亿年左右的史前时代,花朵已经遍布地球的每一个角落。但是,再往前追溯,这些会开花的植物却神秘的消失了,完全找不到它们演化的证据。开花植物究竟是如何出现,又是如何演化的呢?达尔文做了大量调查,却没有找到线索,就将它称为"讨厌的谜"。1998—2002 年,我国古植物学家孙革率领课题组在辽宁西部早白垩世义县组(距今约 1.25 亿年)首次发现了当时世界上最早的被子植物——"辽宁古果"和"中华古果"化石,为花朵的早期演化找到了化石证据。

下面,就请大家随我到左侧展厅来看看"植物与人类"。

植物与人类

○ 第一段

欢迎大家来到"植物与人类"展厅。植物与人类的关系密不可分,从采摘果实到钻木取火、从纺织到建筑,人类的衣食住行都离不开植物。

在这里,我们可以看到原始的野生稻能结出的种子很稀少,而育种之后的水稻可以结出更多的种子。现在,我来问问大家,我们吃的大米是水稻的哪个部分呢?(种子的胚乳)

大家知道我们常吃的"五谷"指的是什么吗?其实,古人说的五谷和今天的五谷不太一样,古人说的五谷,指的是稻、黍、稷、麦、菽。其中,稻和麦大家比较熟悉,稷俗称"糜子",是禾本科黍属植物,是我国北方的主要粮食作物;黍是黄米;菽是豆类。现在的五谷杂粮多指稻谷、麦子、高粱、大豆和玉米。

古代有神农尝百草治疗疾病的传说。当今,我国科学家屠呦呦的团队从植物中提取出青蒿素,挽救了数以百万疟疾患者的生命。在植物身上,有许多值得人类挖掘的宝藏。

中药文化是中国传统医学文化的重要组成部分,起源可追溯数千年。在原始社会,人们就利用野生草药治疗疾病。商周时期中药开始形成体系。这里展示了一些大家耳熟能详的中成药的原材料,如甘草,熟地黄,枇杷叶等。

○ 第二段

油料植物不仅能提供花生油、玉米油、大豆油等食用油,也能提供工业用油,如蓖麻油、桐油等。油桐就是重要的工业油料植物,油桐的种子榨出的是一种优良的干性植物油。桐油用途广泛,是制造油漆、油墨的主要原料。它产油量高,是花生的 5~6 倍,大豆的 10 倍,有"世界油王"之称。

贝叶棕属于棕榈科,是一种常绿乔木,它原产于亚洲的热带地区,是热带地区绿化环境的优良树种。贝叶棕树高约 20 米,树干上没有枝丫,树冠像一把巨伞,叶片像手掌一样散开,看起来笔直浑圆、高大雄伟。

贝叶经就是写在贝树叶子上的经文,有 2500 多年的历史。它源于古印度,具有极高的文物价值,有"佛教熊猫"的美誉。贝叶经是用铁簪子将文字刻写在贝叶上,再涂上颜料,字迹清晰擦抹不掉。再经过水煮等特殊工艺处理,还能防虫、防水、防变形,这也是贝

生命之绿 *The green of life*

叶经可以保存千百年不腐的原因。中国西藏是当今世界保存贝叶经最多、最丰富的地区。同时,贝叶经也是研究古代西藏宗教、艺术等方面重要的原始资料。

○ 第三段

这里是纤维植物的标本。全世界有数百种纤维植物。虽然现在合成纤维在我们的生活中应用十分广泛,但是造纸等行业依然无法离开纤维植物。在合成纤维被发明之前,人们的衣服都离不开植物纤维,和我们关系最密切的就是棉花。这边展示的是一套男士树皮衣,该展品是我国西南地区少数民族以箭毒木(俗称见血封喉)的树皮为原料制作而成的。

箭毒木的乳白色汁液含有剧毒,一经接触人畜伤口,就会使中毒者心脏麻痹、血液凝固,甚至窒息死亡。箭毒木树皮厚、韧性好,树皮中纤维交错缠绕在一起,仿佛编织一样,经过复杂的制作工艺,用它来做衣服,具有防虫耐腐,易穿耐洗的特点,因此备受先民的喜爱。大自然虽然创造了这种剧毒的植物,但是只要善加利用,它们同样可以造福人类。

○ 第四段

香料植物是一类能够产生并可从中提取出芳香物质的植物,我国芳香植物的种类超过 600 种。

提到香料植物,大家一般会联想到香水。除了香水,还有植物精油、薄荷油、清凉油、樟脑等。其实,和我们人类关系最密切的是食用香料植物,比如花椒、大料、八角、桂皮、芥末等等,几乎每天都会出现在人们的餐桌上。

植物的根、茎、叶、花或果实中存在大量的次生代谢产物,其中有些产物可以使其他材料着色,因此被称为染料植物。叶绿素、胡萝卜素、番茄红等天然色素能加工成食品色素,让食物看起来更加美味诱人。同时,植物染料也广泛运用在国画、唐卡、瓷器、蜡染等艺术作品中。

现在,大家再随我到右侧展厅,继续了解关于植物的知识,看一看我们身边的花草,了解"被子植物的繁盛与适应"。

第1章 绿色生命的倾诉
Chapter 1 The Discourse of Green Life

被子植物的繁盛与适应

○ 第一段

　　欢迎大家来到"被子植物的繁荣与适应"展厅。被子植物又称为有花植物,它们的种子大多被包在果实里面。是植物界演化水平最高、种类最多、分布最广的一个类群。现存的被子植物有20多万种,约占植物界的一半,我国就有3万多种。

　　被子植物为什么如此繁荣,它们靠什么样的结构和生存技巧遍及地球上的每一个角落,为世界增添缤纷的色彩呢？现在,让我们一同走进被子植物的世界。

○ 第二段

　　被子植物有根、茎、叶、花、果实和种子六大器官。首先要为大家介绍的是被子植物的叶。叶通过叶绿体进行光合作用,从而产生有机物,叶承担着蒸腾与呼吸的作用。是植物的重要营养器官。少数植物还能通过叶进行繁殖,如我们熟悉的多肉植物中就有很多是落地生根(扦插)。

　　一般植物的叶可分为叶片、叶柄和托叶三部分。若三部分都有,称为"完全叶",比如月季等植物的叶;若缺少叶柄或托叶,称为"不完全叶",比如白菜等植物。不同植物的叶片形态多种多样。

　　在这面展墙上,一共展出了127种被子植物的137件不同形态的腊叶标本。它们的叶形、叶缘、叶尖都不相同,这正是植物为了适应不同的环境,进行多样性演化的最好证据。看到这么多叶子,大家是否想过,为什么有的叶片是绿色,有的则是红色？大多数植物的叶片含叶绿素最多,就会呈现绿色。当秋天来临时,银杏、三角槭等树木的叶片变成黄色;枫树、黄护树等树木的叶片变成红色。因为这些树木的叶片中除含叶绿素外,还含有类胡萝卜素、藻红素或者花青素。

　　叶形相近就代表植物之间有亲缘关系吗？这可不一定。植物的根、茎、叶会随着环境改变外形,不同的植物在相同的环境下生长,经过长时间的演化之后,也可能演变出相近的叶形。相对而言,相同的植物在不同的环境下生长,可能演化出形状、厚薄和大小都不一样的叶子。因此,辨识植物是不是有亲缘关系,必须从花形、花蕊、树木及排列情形等方面判断。

　　了解了被子植物的叶,我们再来看看被子植物的花。通常,花由花冠、花萼、花托、花

生命之绿 *The green of life*

蕊组成。花是被子植物重要的繁殖器官,色彩鲜艳、种类繁多的花,装点着美丽的地球。

在本单元陈列了花的 18 份腊叶标本和 28 份浸制标本。所谓花序,就是花在花轴上不同形式的序列。这面墙以图片和多媒体视频的方式,向大家介绍了花朵传粉的秘密。

这里是植物果实的浸制标本。一般的植物标本都是褪色干瘪的,但是现在用"化学置换保色技术"来制作浸制标本,利用化学分子置换固定植物中各种有颜色的物质,可以让植物保色 10~50 年。

○ 第三段

这里展出了 181 种果实和种子,其中最吸引人的要数象鼻棕的果序了,象鼻棕属于棕榈科。它的形态粗壮下垂,好像大象的鼻子,果实呈椭圆形或倒卵球形。在植物学上,象鼻棕被称作一次花果植物,即一生只开一次花、结一次果,然后就全株枯死。它的寿命约为 20 年,在多年生植物中,一次花果植物并不多见。象鼻棕原产于非洲的热带地区,我国云南的西双版纳、广西南宁以及台湾等地有引种栽培。这个珍贵的巨大标本就是采集自中科院西双版纳植物园。

板状根和绞杀是热带雨林的两种独特现象。板状根由临近地面的粗大侧根发育而来,热带雨林中树木的树形高大,板根的存在增加了树木的支撑能力,可以解决高大乔木"头重脚轻"站不稳的问题。

绞杀是热带雨林中一种常见的现象。例如,桑科榕属的植物,它们的种子会通过鸟类等小动物传播到其他树上,种子发芽长大后,会夺取寄主植物的水分和养分,寄主植物会因外部的压迫和内部养分的匮乏死亡腐烂,而绞杀植物就成了这里"顶天立地"的新主人。植物们为了争夺雨林中珍贵的阳光和有限的生存空间,每天都在进行着无声且残酷的斗争。

望天树又叫擎天树,是中国云南特有的珍稀树种,高度可达 40~60 多米,它们只分布在西双版纳 20 平方千米雨林的范围内,该物种已被列为国家Ⅰ级重点保护野生植物。

○ 第四段

为了适应荒漠严酷的生存环境,荒漠植物大多植株低矮、下半部分枝多,发达的根系有助于吸收土壤水分,细小狭窄的叶片可以减少水分蒸发,这些特征可以帮助荒漠植物有效抵抗荒漠地区的干旱和风沙。

这是内蒙古阿拉善盟的巴特尔先生捐赠的胡杨茎干。胡杨是沙漠中唯一天然成林的乔木,被誉为"沙漠的脊梁""沙漠英雄树",它有抗干旱、御风沙、耐盐碱的能力,是执着坚韧、顽强不屈的象征。

胡杨有一个三千年的美丽传说,据说它可以生而不死一千年、死而不倒一千年、倒而不朽一千年,由此我们可以感受到胡杨顽强的生命力。胡杨又称"异叶杨"。在标本中,大家可以看到,在不同的生长时期、不同的生长部位,同一棵树的叶形可能会不同。为什么胡杨树的叶形有如此大的区别呢?这和它的生长环境有关。沙漠地区极度缺水,幼叶细小,狭长如柳叶一般,这样可以减少光合作用的速率,减少蒸腾,节约水分。等到根部深深地扎入地下,树木吸收到了足够的水分,树叶就会逐渐变大,变为宽卵形或心形,圆润得就像杨树叶片一般了。

○ 第五段

这里,向大家介绍红树林。红树林并不是红颜色的树。因为这些红树科植物主要分布在热带和亚热带的海滩上,所以统称为"红树林"。当台风和海啸来临时,红树林是天然的屏障,为地球家园挡风御浪。因为它们的存在,很多沿海城市躲过了自然灾害,因此,它们被称为"消浪先锋""海岸卫士"。

红树林最奇妙的繁殖方式就是"胎生"。胎生是哺乳动物的特殊生殖方式,简单讲就是妈妈从肚子里直接生出宝宝。那红树林的植物又是如何"胎生"的呢?原来,因为红树林长期经受海潮涨落和海浪冲击,种子很难有一个稳定的萌发环境,所以红树林中很多植物的种子在还没有离开母体时,就已经在果实内部萌芽生长成一个个小幼苗,它们不断地从种子里吸取养料,慢慢变成像"笔"一样的小锥子。等到时机成熟,就会直直落下并插入海滩的淤泥中,并且很快扎根生长为新的植株。

○ 第六段

高寒地区常年低温,生存环境极其严酷,海拔每升高100米,温度大约会下降0.6摄氏度,因此高寒植物一般都贴地生长。囊种草是一种垫状植物,它的直径可达1.3米以上,枝条紧密拥挤在一起,这样就可以阻止热量和水分散失,适应寒冷和大风天气。因此,它们也被称为"垫状植物之王"。

现在,我们来看看"聪明"的植物类群——食虫植物,它们是一类会捕获并消化动物,并从中获得营养的植物,属于植物界中的特殊类群,全世界约有500种。食虫植物能通过光合作用获得养分,它们大多生长在土壤贫瘠,特别是缺少氮素的地区,需要通过捕捉昆虫,来补充植物生长发育所需的营养。

夏天,人们也许会买上几盆猪笼草,希望它能帮助我们"吃点蚊虫"。其实这招并不灵验,因为靠吸血维生的雌蚊对蜜液和紫外线都不感兴趣,所以并不会飞到猪笼草那里自投罗网。我来考考大家,猪笼草袋囊的盖子有什么作用呢?其实,盖子的作用只是装饰而已，根本不会往下盖（顶多遮挡一下雨水，以防消化液被稀释）。昆虫是因为受到捕虫囊口缘的蜜汁吸引，一面吸吮蜜汁，一面走向滑溜溜的捕虫囊内侧，最终自投罗网。

接下来，我们介绍一种比较自私的植物——寄生植物，它们以活的有机体为食，从绿色植物中取得其所需的全部或大部分养分和水分。例如，草苁蓉就不能进行光合作用，它们寄生于赤杨属植物的根部。中医认为草苁蓉能强身滋补，过去经常遭到盗挖。草苁蓉属于国家二级保护植物，人类不应该为了个人欲望破坏地球的生态环境。

听了我的讲解，大家对植物是不是有了一定的了解呢？谢谢各位，我的讲解就到这里。

第 2 章　走近绿色生命的科学家

Chapter 2 Getting Closer to Green Life Scientists

The Scientist Who Approached the Green of Life

Carl Linnaeus, the Father of Biological Taxonomy

(May 23,1707–January 10,1778)

"June 24th.

I thank the Lord for the blessing of this spring and summer bloom, and for this earth that is more perfect than anywhere else –air, water, green spaces, and birdsong! This morning I went out to gather plants."

—— Carl Linnaeus

Students, when it comes to our standard plants, like corn. What do people from different places call corn? Pods, sticks, bracts, yucca shucks, corn, yucca wheat ...... Every plant, and every animal, has its own unique name, or quite a few familiar names and aliases. However, for scientists, each species actually has only one unique Latin scientific name, for example, the Latin scientific name for corn is *Zea mays* L.. So where did this biological name come from?

Well, that's where the great Swedish biologist Carl Linnaeus comes in. It was he who first conceived the principle of defining genera and creating a unified system of biological nomenclature that would provide order to nature in a simple and orderly way.

Carl von Linnaeus was born on May 23, 1707, in a small town in southern Sweden. In Linnaeus' time, most Swedes did not have surnames, and it was only in the generation of Linnaeus' father, Nils, that surnames were given to the family. Because a large linden tree (Lind) grew in the Linnaeus' fields, Nils gave his family a Latin surname: Linnaeus.

Linnaeus' father was a priest and a farmer who loved plants. Under his father's influence, Linnaeus grew very fond of plants. At the age of five, he already began to take care of a small garden, and at the age of eight, he was nicknamed "the little botanist" by his neighbors. The childhood Linnaeus often saw unknown plants to ask his father, and his father also told him in detail. Sometimes, when Linnaeus couldn't remember all his father's answers and asked questions again, his father would urge Linnaeus to strengthen his memory by "not answering the questions he had asked", which made his memory exercise since he was a child. Although the young Linnaeus knew more and more kinds of plants, he did not excel in his studies but only had an extraordinary love for various plants.

In 1727, Linnaeus enrolled at the University of Lund in Sweden, and a year later, he transferred to a better school, the famous Uppsala University in Sweden. During his time at the university, he made full use of the university library and botanical gardens for museum studies, and at the same time, systematically learned the knowledge and methods of collecting and making biological specimens.

In 1729, Linnaeus read the book The Structure of Flowers and Plants written by the French botanist Villante and thus was inspired to start classifying plants according to the number of pistils and stamens.

In 1732, Linnaeus joined an expedition financed by the Academy of Sciences in Uppsala, and traveled with the expedition to the Lapland region in northern Sweden, located within the Arctic Circle, for a field trip. At that time, Europeans knew nothing about the Lapland region, and in this vast and inhospitable area, the young botanist of the future, with simple traveling clothes, trekked 2,500 kilometers in six months, far beyond the Arctic Circle. The Lapland region is full of rivers and swamps. Linnaeus' only means of transportation were his own legs, a horse, and a boat. It was strong faith and intense curiosity that turned the young man's dangers into success time and time again. In the large and small swamp area, sometimes the cold sewage will flood his stomach; the rugged road, so that he fell off the horse several times; those like dark clouds of mosquitoes, dense, chasing him; barren areas of food scarcity, often only maggot-covered dried fish, reindeer meat and milk …… However, Linnaeus remained fully absorbed in observing the soil and the flora and fauna.

The further north Linnaeus traveled, the sparser the vegetation became. He carefully searched the caves and risked his life to collect many kinds of mosses, lichens, and ferns in them. In the Alps, located in Lapland, Linnaeus struggled to climb in snowstorms. He wrote in his diary, "I was lost in wonder and thought to myself that I could have found more, I just didn't know what to expect." During this time, Linnaeus found a shrub. This small shrub was inconspicuous in the shady coniferous forest, but Linnaeus was so in love with it that he not only made him his symbol but simply named it, after himself, lin-naea, which means Linnaeus wood. Chinese translates it as Arctic Flower. This flower mostly grows around the North Pole, but it is also found at high altitudes in the Changbai Mountain range in northeast China, far from the North Pole. 6 months of exploration allowed the young Linnaeus to discover more than 100 kinds of new plants and collect a lot of valuable information.

In 1732, Linnaeus returned to the University of Uppsala, where he published the results of his investigations in his work "Flora of Lapland". At this time, in his mind, he was always organizing, classifying, and comparing these plants, and he was determined to come up with a clear order for these marvelous gifts of nature.

In 1735, Linnaeus traveled around Europe. There, he met his good friend P. Artedi. Both of them were dissatisfied with the classification system of living things at that time, and they aspired to make a brand-new classification system together, with Linnaeus focusing on plants and Artedi on animals. They also made a promise to each other that if one of them died first, the other would finish the other's unfinished work. Who knew that the prophecy would come true? Not long afterward, Artedi drowned because of an accident.

The tour of European countries was the most important period of Linnaeus's life, and also the stage of maturity and the first appearance of his academic ideas. 1735, Linnaeus published a thin pamphlet, Systema Naturae (System of Nature). Later, this work was reprinted again and again, with an ever-expanding length, and by the time Linnaeus died, it had already been published in its 12th edition, becoming a huge book of more than 2,000 pages. In this book, Linnaeus not only proposed a brand-new classification system for plants, but also a brand-new classification system for animals, fulfilling his promise with Arthédi, which was really a scientific friendship across life.

In 1738, Linnaeus returned to Sweden and in 1741 became a professor at his alma-mater, Uppsala University. Since then, Linnaeus has been engaged in teaching and research and has trained many students.

During the period in which Linnaeus lived, it was the age of great European exploration. Many naturalists and biologists returning from voyages brought back specimens of various organisms from around the world. They often named these organisms based on their personal preferences, resulting in confusion with multiple names for one species or the same name for different species. Furthermore, due to the different languages and writings of scholars from various countries, the names of organisms in the field of botany, for example, often became exceedingly long.

In 1753, Linnaeus published "Species Plantarum," introducing the binomial nomenclature system and using Latin to name living organisms, including plants. Latin was the universal language of the ancient Roman Empire, although by Linnaeus' time in the 18th century, it was rarely used in everyday life. Due to its limited use, Latin had undergone minimal changes and development in this era. Linnaeus classified plants based on characteristics such as the types, sizes, quantities, and arrangement of their stamens and pistils. He divided plants into 24 classes (class), 116 orders (order), over 1,000 genera (genus), and more than 10,000 species (species). Linnaeus was the pioneer of the class (class), order (order), genus (genus), species (species) classification system. He introduced binomial nomenclature, where the common name of a plant consists of two words: the first word is the "genus name," and the second word is the "species epithet." If needed, the initial letters of the name of the person who discovered or named the species can be included as an abbreviation in honor. For example, the scientific name of the ginkgo tree is Ginkgo biloba L., where Ginkgo is the genus name, which is a noun, and biloba is the species epithet, which is an adjective. The final letter, L, is the abbreviation of Linnaeus. To keep names concise, Linnaeus specified that a scientific name should be limited to 12 words, making information clear and easy to organize. Later, he applied the same naming system to animals, and this nomenclature is still in use today.

On January 10, 1778, this great Swedish biologist, who had reorganized the natural order of the biological world, Carl Linnaeus, passed away. Linnaeus' collection of books and specimens was sold to the British, and the British government established the Linnean Society in London in 1788. His manuscripts and biological specimens were preserved there. Even today, the Linnean Society is a renowned scientific society worldwide. To commemorate this outstanding scientist, the Swedish government has established institutions such as the Linnaeus Museum and the Linnaeus Botanical Garden, and in 1917, the Linnaeus Society of Sweden was founded. Within the prestigious institution, the University of Chicago, there stands a statue of Linnaeus, and many places around the world commemorate and remember this scientist in various ways.

When we stand in the National Museum of Natural History, we can look up at the large portrait of Linnaeus hanging in the exhibition hall. His resolute lips seem to be telling us about the natural order of all things in the world⋯

Chinese Pioneer in Plant Taxonomy——Hu Xiansu

(April 20,1894—July 16,1968)

In the National Museum of Natural History's collection of fine specimens, there is an ancient specimen of a tulip tree. The tulip tree, native to Europe, is a famous ornamental plant. This precious specimen was collected, prepared, and identified by Hu Xiansu, the founder of Chinese plant taxonomy, in Nanjing, Jiangsu, in 1920. At that time, social unrest prevailed, and Hu Xiansu and others traveled through Zhejiang, Jiangxi, Fujian, and other regions, collecting a large number of plant specimens. He was the first person to conduct botanical fieldwork within Jiangxi's borders. Based on this collection, Hu Xiansu published a series of articles, including "A List of Plants in Zhejiang" and "Miscellaneous Notes on Fungi Collected in Jiangxi." This specimen, as a witness to history, was originally stored in the plant specimen room of the Institute of Botany, Beijing Academy of Sciences. It has since been transferred to our museum. After more than 100 years, the specimen is still well preserved.

Hu Xiansu was born in Nanchang, Jiangxi, and was known as a child prodigy due to his exceptional intelligence, earning him the nickname "Child Genius." This young prodigy entered the preparatory department of the Imperial University of Peking (now Peking University) at the age of 15 and was sent to study in the United States on a government scholarship at the age of 19. During his time in the United States, he obtained a bachelor's degree from the University of California and later earned a Ph.D. in plant taxonomy from Harvard University.

After his first return from the United States, he taught at various universities in China, including Nanjing Higher Normal School, Southeast University, Peking University, and Tsinghua University. Hu Xiansu, based on the domestic situation and his education in the United States, co-authored the Chinese textbook "Advanced Botany" along with Qian Chongsu and Zou Bingwen. This was China's first university-level biology textbook.

In 1925, Hu Xiansu founded the earliest biological academic publication in China, the Series of the Institute of Biology of the Chinese Society of Sciences, to strengthen academic exchanges.

In 1928, Hu Xiansu with Bingzhi and others founded the Institute of Biology under the China Science Society and the Institute of Cryptogamic Botany under the North China Institute of Botany, serving as the head of the plant department. At the time, the field of biology in China was just beginning to take shape. The North China Institute of Botany witnessed the history of the development of biology in China from its inception.

Hu Xiansu attached great importance to the development of plant taxonomy in China and advocated an educational philosophy of "using science to save the nation, practical application of knowledge, self-reliance, and independence." He emphasized talent cultivation, organized field surveys, collected specimens, and promoted international exchanges. He also initiated the preparation of the Flora of China, a comprehensive reference work on Chinese plants.

In 1933, Hu Xiansu established the Botanical Society of China and founded the first botanical garden in China, Lu Shan Botanical Garden. He was also the founder of the herbarium at the Institute of Botany, Chinese Academy of Sciences.

In 1946, Hu Xiansu received some unique plant specimens, including branches, leaves, spherical flowers, and young fruit capsules, from friends. Based on these specimens and extensive literature research, he identified and named a new plant species – the Dawn Redwood. Prior to this discovery, it was widely believed in the botanical community that the Dawn Redwood was an extinct species. Therefore, the discovery and naming of the Dawn Redwood astounded botanists worldwide. In 1948, the Dawn Redwood was successfully introduced and cultivated in Lu Shan Botanical Garden. This rare plant, similar to the Giant panda in rarity, was subsequently introduced and planted in various parts of the world. Hu Xiansu became known as the "Father of Modern Dawn Redwoods."

Chairman Mao Zedong praised Hu Xiansu as the "forefather of Chinese biology." In

1962, Hu Xiansu composed a classical scientific poem in ancient style titled "Song of the Dawn Redwood." This poem was published in the People's Daily, and Vice Premier Chen Yi highly praised it, writing his thoughts after reading it.

Hu Xiansu proposed and published a "Multiple-system approach to the classification of angiosperms" and a phylogenetic system diagram of angiosperms, which were established by Chinese plant taxonomists for the first time. Throughout his life, he authored more than 20 monographs, published over 140 botanical papers, and discovered one new family, six new genera, and over 100 new species of plants.

Today, the Dawn Redwoods at Lu Shan Botanical Garden stands tall and luxuriant. Alongside the Dawn Redwood, future generations remember the pioneer of plant taxonomy, Academician Hu Xiansu.

In a tumultuous era, he adhered to his original aspiration, and promoted the establishment of Chinese plant biology, and, to this day, the development of Chinese plant biology stands as tall and flourishing as the Dawn Redwoods.

The "Living Dictionary" of Chinese Plants——Wu Zhengyi
(June 13,1916—June 20,2013)

The Kunming Institute of Botany at the Chinese Academy of Sciences has a stone monument engraved with the words "原本山川,极命草木" ("To understand the origins of mountains, rivers, and the essence of plants"). In 1938, Hu Xiansu, the director of the North China Institute of Botany, and Gong Zizhi, the director of the Yunnan Provincial Department of Education at the time, decided to establish the Yunnan Institute of Agriculture and Forestry (predecessor of the Kunming Institute of Botany, Chinese Academy of Sciences). They proposed to use these eight characters as the institute's motto. Gong Zizhi personally wrote and engraved these characters on a stone tablet, which was mounted on the institute's wall, but unfortunately, it was later destroyed.

The phrase "原本山川,极命草木" is derived from the "七发" (Seven Disputes), a famous work by the Western Han Dynasty poet and essayist Mei Cheng. It means to organize and summarize the names of mountains, rivers, plants, and trees, making them part of literature and preserving them for posterity. This motto serves as the foundation of the Kunming Institute of Botany and represents the philosophy and spirit of Chinese botanists. In his later years, Academician Wu Zhengyi rewrote these eight characters and had them engraved in stone. "原本山川,极命草木" epitomizes Wu Zhengyi's lifelong connection with plants and serves as a vivid portrayal of his exploration of Yunnan.

Wu Zhengyi's family was originally from Anhui's Shexian County, but he was born in Jiujiang, Jiangxi, and his family relocated to Yangzhou, Jiangsu, when he was one year old. Even as a child, Wu Zhengyi exhibited a strong interest in plants. He observed the changing seasons, and the blooming and withering of flowers in his own backyard, and this sparked his interest in botany. During his youth, his favorite readings included "植物名实图考" (A Comprehensive Study of Names and Facts of Plants) by Wu Qijun from the Qing Dynasty and "日本植物图鉴" (An Illustrated Guide to Japanese Plants) by the modern Japanese

botanist Makino Tomitaro.

After completing his secondary education, Wu Zhengyi decided to enroll in the Department of Biology at Tsinghua University, despite his father's objections. At the time, China was in turmoil, and many young people chose to study abroad or pursue science majors at top domestic universities. Wu's father believed that studying plants and flowers was a useless endeavor. However, the young Wu Zhengyi firmly stated, "I love plants and want to excel in their study."

In 1937, Wu Zhengyi graduated from Tsinghua University with a degree in biology and stayed on as a faculty member. That same year, as the Second Sino-Japanese War broke out, Tsinghua University, Peking University, and Nankai University formed a temporary university in Changsha. In late 1937, the temporary university moved to Kunming, Yunnan, and became known as "Southwest Associated University." Wu Zhengyi, along with other professors like Wen Yiduo and Li Jie, led students on a 68-day, 1663.8-kilometer journey on foot from Changsha to Kunming.

During the turbulent wartime years, Wu Zhengyi actively participated in the anti-Japanese resistance. He joined the Chinese Communist Party in 1946. Despite challenging conditions, he continued to collect botanical books, conduct field surveys, and organize and write detailed botanical cards, totaling nearly 30,000 cards. Each card contained comprehensive information on the scientific names, distribution locations, and relevant literature for various plants. These cards became the foundational data for Wu Zhengyi's later compilation of the "Flora of China."

Within the rudimentary herbarium he assembled from wooden oil crates, Wu Zhengyi systematically organized and identified many specimens that had not yet been mounted on herbarium sheets. This pioneering effort marked the beginning of Chinese botanists' independent identification of plant specimens. He personally wrote, illustrated, and printed "滇南本草图谱" (Illustrated Flora of Southern Yunnan), laying the groundwork for the emerging field of Chinese botanical philology.

After the victory in the War of Resistance Against Japan, Wu Zhengyi served as the first secretary of the Party branch at the Chinese Academy of Sciences and as a researcher and deputy director at the Institute of Botany, Chinese Academy of Sciences. In 1955, he was elected as one of the first members of the Chinese Academy of Sciences. However, Wu Zhengyi began to contemplate his scientific journey and life path in earnest. He longed for

Yunnan's tropical seasonal rainforests, montane evergreen broad-leaved forests, alpine meadows, subalpine coniferous forests, and various secondary vegetation types. He recalled the grand vision he had set for himself in his youth: "To establish myself in Yunnan and survey the plants of China and the world."

In the scorching summer of 1958, at the age of 42, Wu Zhengyi, along with his wife and young children, made the resolute decision to relocate to Yunnan and initiate the establishment of the Kunming Institute of Botany, embarking on a profound connection with Yunnan that would last for nearly half a century. Starting from Yunnan, Wu Zhengyi's footprints extended across the mountains and valleys of China, and he explored plant diversity on four continents outside of Africa. Wherever there were plants, there was Wu Zhengyi's indomitable spirit of exploration.

Wu Zhengyi is recognized as one of the Chinese botanists who discovered and named the most plant species, fundamentally altering the history of Chinese botany that had been dominated by foreign scholars. He named or participated in naming a total of 1,766 plant taxa, encompassing 94 families and 334 genera, including 22 new genera. Three plant species bear his name, including the Chinese endemics "Zhengyi Osmanthus" and "Zhengyi Cotoneaster," as well as "Zhengyi Hemp" discovered in Shennongjia, Hubei.

Wu Zhengyi devoted more than 70 years to plant research and teaching, making groundbreaking contributions to understanding the origins of Chinese plants. He was the first to propose the three historical sources and 15 geographical components of the Chinese plant flora. Did you know that understanding the flora of a particular region entails knowledge of its origins, formation, evolution, and development? Investigating the origins of Chinese plants is like tracing the intricate history of these plants.

Wu Zhengyi led the compilation of the "Flora of China." This monumental work consists of 80 volumes (126 fascicles), two-thirds of which were completed under his leadership as the chief editor. Additionally, he served as the chief editor of the "Flora of Tibet" and "Flora of Yunnan," among other publications. His efforts facilitated the work of later botanists by providing convenient resources for the study and identification of plants. This is a complex and massive project. The more than 30000 plants in the book, most of which grow in the vast mountains, with each flower and each grass, are precious plant resources that Wu Zhengyi collected after conducting field investigations.

Furthermore, Wu Zhengyi made significant contributions to the rational utilization and conservation of China's plant resources. As early as 1958, he submitted proposals for the planning and establishment of 24 natural reserves to the Communist Party Committee and the Government of Yunnan Province. His recommendations, such as "creating a 'Hua Guoshan' to transform barren mountains into economic forests," were adopted by various local governments. He provided numerous effective suggestions for the specific utilization and protection of plant resources. His proposal to establish a national repository for wildlife germplasm resources is now being implemented as a major scientific project.

Wu Zhengyi's life was marked by unwavering dedication to his research, regardless of the time or place. He understood that only by achieving outstanding results could he maximize the value of his field to his country. During special periods, he demonstrated an astonishing memory by personally writing over 500,000 characters for the "New Flora of China," which included more than 6,000 medicinal plant species and boasted an accuracy rate of over 95%. He truly earned his title as the "Living Dictionary" of Chinese plants.

His footprints spanned north and south, and his writings included numerous botanical works that are considered classics by the plant science community. He is regarded as an unassailable peak in the field of botanical research in China and the world. On December 10, 2011, the International Astronomical Union's Minor Planet Center permanently named asteroid number 175718 "Wu Zhengyi Star." From the universe to the stars, in the vast expanse of the sky, he continues to unravel the mysteries of nature and the origins of plants.

The "Father of Hybrid Rice"——Yuan Longping

(September 7,1930—May 22,2021)

I have a dream. Buried deep in the soil; I believe its unique. Light gives it hope. Rain provides nourishment. It accompanies seeds as they grow.

I have a dream. Walking along the field path. It stands as tall as me. I'm with my dearest friend. Sitting under the rice stalks, seeking shade.

Mother, I've come to see you. Look at the evening glow over the village. Mother, I'm here to talk to you. These seeds were sown by your hands. And they've sprouted in my heart.

The wind rustles the rice waves. Rice husks brush against my palms. Rice straw is piled into heaps in the field. Grains glisten in the sunlight. The paddy fields turn golden orange.

Whenever we sing this song, we think of China's "Father of Hybrid Rice," Yuan Longping. "I have a dream" not only expresses Yuan Longping's dream of "finding shade under the rice," but also his deep love for his mother. On May 22, 2021, Yuan Longping left us forever. He once said, "A person is like a seed; to be a good seed." He thought of the nation and its land and left for the rice fields. Yuan Longping changed our world and has gone to another world to fulfill his dream of "rice taller than people, where one can find shade."

His entire life was dedicated to our motherland and to rice. He consistently set new records for crop yields. One of the ways we can remember Yuan Longping is by cherishing every meal, as each grain of rice is hard-won.

There is much knowledge about hybrid rice that many students might not be familiar with. Is hybrid rice a genetically modified crop? What is the difference between hybrid and genetically modified crops?

Hybridization refers to the process of crossing two parent plants, typically one male and one female, in the breeding of rice. Hybrid rice is not a genetically modified crop; instead, it involves selecting two rice varieties with certain genetic differences and complementary desirable traits, crossing them, and producing hybrid seeds that exhibit hybrid vigor. These hybrid seeds are then used for cultivation.

Genetically modified crops, on the other hand, involve the insertion of specific genes into a plant's genome to introduce new traits or modify existing ones. This is typically done through advanced biotechnology methods.

For hybrid rice, what has been artificially cultivated and preserved is breeding: if there are any shortcomings in the hybrid variety, it will be eliminated and other varieties will be reselected. Although genetically modified soybeans have the same appearance as soybeans, they are no longer completely the original "soybeans".

For hybrid rice, the fundamental idea, and techniques, as well as its first successful implementation, were achieved by American scientist Henry Beachell in Indonesia in 1963. Consequently, Henry Beachell is also known as the "Father of Hybrid Rice" and was awarded the World Food Prize in 1996. However, his approach had certain limitations that hindered large-scale adoption.

Subsequently, Japanese scholars proposed the three-line breeding method to develop hybrid rice, using suitable wild male-sterile lines as the foundation. Although many years of effort were invested, the results were not very promising. In China, over 60% of the population relies on rice as their main food source. In the southern regions, more than 80% of the population depends on rice. The rice yield plays a decisive role in ensuring food supply and security.

In the early days of the People's Republic of China, the average yield of rice per acre was less than 150 kilograms, and food was extremely scarce. Hybrid rice was successfully researched in 1973 and began large-scale promotion in 1976, gradually addressing food shortages. Today, the national average yield of hybrid rice is nearly 550 kilograms per acre, with the highest yield reaching 1,149 kilograms per acre. The increased food production from planting hybrid rice each year can feed an additional 80 million people. Yuan Longping, using less than 9% of the world's arable land, is feeding 20% of the world's population, creating a miracle in the history of global food production.

Yuan Longping's dream was to change the world with a single seed. He dedicated his entire life to this dream. Every scientific experiment required long waiting periods—half a year, one year, two years—without shortcuts or acceleration. Day by day, year by year, he persevered like a seed, taking root in the soil, unafraid of storms. Through diligent research and steadfast effort, he eventually realized this great dream. Therefore, he is also known as the "Father of Hybrid Rice" in China.

Over the past 40 years, the scope of knowledge transfer for hybrid rice technology has continuously expanded, taking root, and flourishing in many countries. Hybrid rice is now grown in over 40 countries. In Madagascar, Africa, hybrid rice yields are more than double those of local varieties. To commemorate this great achievement, the country has printed the image of Chinese hybrid rice on its new currency.

China has about 1.5 billion mu (approximately 100 million hectares) of saline-alkali land, accounting for roughly one-tenth of the world's total saline-alkali land area. Approximately 200 million mu (about 13 million hectares) of this land is suitable for growing saltwater-tolerant rice. Yuan Longping and his research team explored maximizing rice production by cultivating saltwater-tolerant rice in these areas. After successful trials, saltwater-tolerant rice yields can reach over 600 kilograms per mu (about 667 square meters). If calculated based on a conservative yield of 300 kilograms per mu, this would yield about 30 billion kilograms of food each year. To put this into perspective, 30 billion kilograms is roughly equivalent to the total annual grain production of Hunan Province.

Under the scorching sun, warm soil, and Yuan Longping's efforts, barren land has transformed into fertile fields. This is no longer a dream.

The Power of a Small Grass——Tu Youyou
(December 30,1930—Present)

On October 5, 2015, the Nobel Prize in Physiology or Medicine was awarded to researcher Tu Youyou from the China Academy of Traditional Chinese Medicine. The Nobel scientific award is the highest honor ever achieved in the field of Chinese medicine and also the highest recognition for an achievement in traditional Chinese medicine. Tu Youyou and her team achieved outstanding success in malaria treatment research, making her the first native scientist in China to win a top award in the natural sciences.

Tu Youyou was born in Ningbo, Zhejiang, in 1930. Her father named her "Youyou," which is derived from a poem in the Classic of Poetry (Shi Jing) that means "the call of a deer." The name "Youyou" seems to have a unique connection with artemisinin, as if they were linked in some strange way over two thousand years ago.

At the age of 16, she had to suspend her education for two years due to a lung tuberculosis infection. From that time on, Tu Youyou developed a strong interest in medicine. In 1951, she enrolled in the School of Pharmacy at Peking University. In her professional courses, she developed a keen interest in plant chemistry, pharmacognosy, and plant taxonomy. This less popular major ultimately connected her with traditional Chinese medicine and malaria treatment.

In 1956, during the nationwide campaign against schistosomiasis, after graduating from university, Tu Youyou conducted pharmacological studies on the effective medicine "Ban Bian Lian"(*Lobelia chinensis* Lour.). Later, she completed pharmacological studies on the more complex Chinese medicine "Yin Chai Hu" (*Stellaria dichotoma* Linn. *var. lanceolata* Bge.). These two achievements were successively included in the "Chinese Pharmacopeia."

In January 1969, Tu Youyou led a research team that systematically collected and organized ancient medical books, Materia medica, and folk prescriptions. Based on more than 2,000 prescriptions, they compiled the "Collection of Antimalarial Single Drug Prescriptions," primarily consisting of 640 drugs, and conducted experimental research on over 200 Chinese herbal medicines. After more than 380 failures and persistent efforts using modern medical methods and improved extraction techniques in extremely challenging research conditions, Tu Youyou and three other researchers eventually discovered and extracted artemisinin. This groundbreaking discovery created a new approach to malaria treatment, benefiting hundreds of millions of people worldwide and saving millions of lives.

However, did you know that the plant Tu Youyou used to extract artemisinin is not true wormwood (Qinghao) but sweet wormwood (Huanghua)?

Tu Youyou explained this as early as 1987. Qinghao and Huanghua are two different Chinese medicinal names. True wormwood (Qinghao) was documented in the "Shennong Ben Cao Jing" during the Han Dynasty, but due to a lack of modern botanical classification and research methods in ancient times, confusion arose during the Ming Dynasty when Li Shizhen's "Ben Cao Gang Mu" was written.

In the 18th century, Japanese botanist Zhang Guan Lidai attributed the Latin name for true wormwood (Qinghao) to another plant belonging to the Artemisia genus that is closely related to the artemisinin-producing plant. Because China imported many botanical works from Japan in modern times, some scholars were influenced by Japanese botanists and accepted this misidentification. Thus, the confusion regarding the name "Qinghao" for the artemisinin-producing plant persisted.

When compiling the authoritative work "Flora of China," researchers, considering that the use of the name "Qinghao" in the botanical world had become a convention, did not change it. They continued to use the incorrect name assigned by Japanese scholars, applying the name "Qinghao" to a plant that had nothing to do with the true wormwood described in ancient Chinese pharmacopeias. They assigned the name "Huanghua" to the artemisinin-producing plant.

In summary, the plant used in traditional Chinese medicine as "Qinghao" is actually sweet wormwood (Huanghua). Sweet wormwood contains artemisinin and can treat malaria, and it has been used in traditional medicine for a long time. The plant referred to as "Qinghao" in botanical studies, while closely related to sweet wormwood, does not possess artemisinin and lacks anti-malarial properties.

The main differences between Qinghao and Huanghua are as follows:

Qinghao often grows on low altitude, moist riverbanks, etc., with plants standing 1-2 meters tall; The lower leaves of the stem are mostly 2-3 pectinate pinnate divisions, while the middle leaves are mostly 2 pectinate pinnate divisions; The inflorescence is

hemispherical in shape.

Huanghua has strong adaptability and is distributed almost throughout the country. The plant is about 2 meters tall, and the lower leaves of the stem are mostly 3-4 pinnate divisions, while the middle leaves are mostly 2-3 pinnate divisions; The inflorescence is approximately an elliptical sphere.

Tu Youyou, despite not being a botanist, should be introduced to students. She is among the outstanding scientists who have made profound contributions to humanity's life and health. Tu Youyou stated that artemisinin is a gift from traditional Chinese medicine to people worldwide.

Plant herbal medicine has opened a door to the world for researchers. Over several decades, Chinese scientists have never stopped exploring...

The Botanist Next Door——Chen Shaoxing

(December 7,1923—July 19,2016)

Today, as we stroll through the colorful botanical exhibition hall, marveling at the enormous brown algae specimen that is as old as the National Museum of Nature, we should not only appreciate the wonders of nature and the passage of time but also remember the dedicated botanist behind these specimens, Chen Shaoxing, the former director of the Botany Department at the Beijing Museum of Natural History.

Behind every exhibit in the museum lies the silent research and dedication of scientists. Chen Shaoxing was born in December 1923 into a merchant family in Shexian, Anhui Province, China. He began his education at a private school at the age of seven, later entering the Shexian County First Elementary School. In June 1935, he graduated as the second in his class and was admitted to Huizhou Middle School. Even in his old age, Chen Shaoxing could recite poems from his youth, like "Thirty years ago, I studied diligently. What makes a hero? In the world, all things must be done. There is no fortress that cannot be conquered." He also received solid English language foundations from his math teacher, Feng Zhiyuan, who used English textbooks and taught mathematics in English.

During the Sino-Japanese War, Chen Shaoxing faced numerous challenges but managed to be admitted to Fudan University's Department of Biology in 1946, ranking fifth

1954年8月,在故宫文华殿举办"祖国自然环境与矿产资源展"。此后陆续举办"农产资源展"等展览。

among more than ten thousand candidates.

In 1950, after graduating from university, Chen Shaoxing worked as an intern editor at the Science Popularization Bureau of the Ministry of Culture. In 1951, he was transferred to the preparatory office of the Central Museum of Natural History to participate in its establishment. During his tenure at the preparatory office, Chen Shaoxing was involved in designing outlines and producing exhibitions, including "Exhibition on China's Natural Resources and Mineral Resources," "Agricultural Resources Exhibition," and "Taiwan Exhibition." These exhibitions were among the earliest held by the museum and were highly successful, receiving enthusiastic responses from visitors, including several national leaders who visited.

In 1959, when the new building of the Beijing Museum of Natural History, the Tianqiao New Hall, was completed and opened to the public, Chen Shaoxing took charge of the Botany Research Department. He led the department in designing botanical displays that emphasized the systematic evolution of the plant kingdom. The plant exhibit consisted of three main parts: higher plant organs and their developmental trends, the evolution of the plant kingdom, and the transformation and utilization of plants. In the autumn of 1959, the plant exhibition opened as scheduled, becoming the first botanical exhibition hall in New China.

In 1962, Chen Shaoxing was responsible for the design of the second botanical exhibition hall, the "Chinese Vegetation Hall," and, along with colleagues from the Botany Department, collected plant

祖国自然环境与矿产资源展

specimens from various parts of the country. The scenes in the exhibition were created using specimens collected and assembled from their original habitats.

In 1998, at an advanced age, Chen Shaoxing participated in the third revision of the Beijing Museum of Natural History's plant exhibition. This reconstruction placed greater emphasis on principles such as species diversity, genetic diversity, and ecological diversity. The newly designed plant exhibit allowed visitors to wander through a tropical rainforest ecological landscape, which was a novel exhibition method at the time.

Chen Shaoxing served as the director of the Botany Research Department at the Beijing Museum of Natural History for an extended period. He also held various social positions, including Vice Director of the Institute of Natural History at the Beijing Museum of Natural History, Deputy Secretary-General of the Beijing Botanical Society, and Director of the China Botanical Society.

Over several decades, Chen Shaoxing conducted numerous plant specimen collection expeditions throughout China and authored dozens of Chinese and English books and papers. Under his leadership, the plant specimen collection at the museum grew from nothing to over 36,000 specimens by the late 1980s.

During an interview with the media, Chen Shaoxing was once asked what spirit had guided him to achieve such abundant professional accomplishments. He replied, "I am just a humble 'Huizhou camel'!" He explained, "The so-called 'Huizhou camel' refers to its ability to endure hardship and fatigue, just like a group of camels traveling across the boundless desert, persistently moving forward. I've been to the Northwest, crossed the Tian Shan Mountains, and ridden camels. In the desert, regardless of wind and rain, with open air and open skies, I moved forward, always moving forward!" Perhaps each of us, in our pursuit of an upward life path, has a little of the spirit of the "Huizhou camel," no matter the adversity or challenge we face.

生命之绿 The green of life

走近绿色生命的科学家

生物分类学之父——林奈

(1707年5月23日—1778年1月10日)

"6月24日

感谢主的赐福,赐予我这春夏盛景,赐予我这片比任何地方都要更加完美的大地——空气、水、绿地、还有鸟鸣!今天早上我出门来采集植物。"

卡尔·冯·林奈

同学们,说起我们常见的植物,比如玉米。来自不同地方的大家,都叫玉米什么名字呢?苞米、棒子、苞谷、玉蜀黍、粟米、玉麦……每一种植物,每一个动物,都有自己独特的名字,或者还有不少俗名和别名。但是,对于科学家来说,每一个物种其实都只有一个独一无二的拉丁文学名,比如,玉米的拉丁文学名是 Zea mays L.。那么,这个生物学学名是怎么来的呢?

这就得说起一位伟大的瑞典生物学家卡尔·冯·林奈(Carl Linnaeus)。是他首先构想出定义生物属种的原则,创造出统一的生物命名系统,用一种简约有序的方式为大自然构建秩序。

卡尔·冯·林奈于1707年5月23日出生在瑞典南部的一个小镇。在林奈的时代,大多数瑞典人是没有姓氏的,到林奈的父亲尼尔斯(Nils)这一辈才为家族取姓。因为林奈家的田园中长着一棵大椴树(Lind),于是尼尔斯便给自己的家族起了一个拉丁语的姓氏:林奈(Linnaeus)。

林奈的父亲是一位牧师,也是一位热爱植物的农夫。在父亲的熏陶下,林奈从小十分喜欢植物。在他5岁时,已经开始照管一个小花园了,8岁时就被邻居们戏称为"小植物学家"。童年的林奈经常将所见到的不认识的植物拿来询问父亲,父亲也一一详尽地告诉他。有时候,当少年林奈不能全部记住父亲的答案而再次提问时,父亲就会以"不答复问过的问题"来督促林奈加强记忆,这使得他的记忆力从小就得到了锻炼。虽然少年林奈认识的植物种类越来越多,但他的学习成绩并不突出,只是对各种植物有着异乎寻常的爱好。

1727年,林奈进入瑞典龙德大学读书,一年以后,他转到更好的学校,瑞典著名的乌普萨拉大学学习。在大学期间,他充分利用大学图书馆和植物园进行博物学的学习,同时,也系统学习了采集和制作生物标本的知识和方法。

1729年,林奈读到法国植物学家维朗特创作的《花草的结构》一书,从而受到启发,他开始根据植物的雌蕊和雄蕊的数目进行植物分类。

1732年,林奈参加了乌普萨拉科学院资助的考察队,和考察队来到瑞典北部位于北极圈内的拉普兰地区,进行野外考察。当时,欧洲人对拉普兰地区一无所知,在这片巨大的荒凉地带,这位年轻的未来植物学家,带着简单的行装,在6个月的时间里跋涉了2500千米,这远远超出了北极圈的范围。拉普兰地区河流纵横,沼泽遍布。林奈的交通工具只有自己的双腿、马和船。正是因为坚定的信仰和强烈的好奇心,让这名年轻人一次次化险为夷。在大大小小的沼泽地区,有时冰冷的污水会淹没他的肚子;崎岖的道路,让他好几次从马上摔了下来;那些像乌云般密集的蚊子,对着他围追堵截;贫瘠的地区食物匮乏,常常只有长满蛆的鱼干、驯鹿肉和牛奶……但是,林奈依然全神贯注地观察着土壤和动植物。

林奈越往北走,植物就越稀少。他仔细搜索着岩洞,冒着生命危险在洞里采集到了很多种苔藓、地衣和蕨类植物。在位于拉普兰境内的阿尔卑斯山脉,林奈在暴风雪中艰难地攀爬。他在日记中写到:"我沉浸在惊奇之中,心想我本可以找到更多,只是我不知道要怎么做。"期间,林奈找到了一种灌木。这种小灌木在阴暗的针叶林中很不显眼,林奈却对它爱不释手,不仅把他当作自己的标志,还干脆用自己的名字,为它命名为 lin-naea,意思是林奈木。汉语把它翻译成北极花。这种花多生长在北极周边,但在远离北极的我国东北长白山山脉的高海拔地区也有分布。6个月的探索,让年轻的林奈发现了100多种新植物,并收集了不少宝贵的资料。

1732年,林奈又回到了乌普萨拉大学,他把调查结果发表在著作《拉普兰植物志》中。这时候,他在心里一直在整理分类和比较着这些植物,他下定决心,要为这些大自然神奇的馈赠排出一个清晰的秩序。

1735年,林奈周游欧洲。在当地,他结识了好友阿尔泰迪(P. Artedi)。两个人都对当时的生物分类系统不满意,他们立志要在一起做出全新的分类系统,林奈主攻植物,阿尔泰迪主攻动物。他们还互相约定,如果有一个人先去世,另一个人要完成对方未完的工作。谁知一语成谶,没过多久,阿尔泰迪就因为一次事故而溺亡。

欧洲各国的游历,是林奈一生中最重要的时期,也是他学术思想成熟、初露锋芒的阶段。1735年,林奈出版了一本薄薄的小册子《自然系统》(*Systema Naturae*)。后来,这本著作一版再版,篇幅不断扩大,到林奈去世之前,已经出版到第12版,成为一本2000多

生命之绿 The green of life

页的巨著。在这本书中,林奈不仅提出了全新的植物分类系统,还提出了全新的动物分类系统,完成了他和阿尔泰迪的约定,这真是一段跨越生命的科学友谊。

1738年,林奈回到瑞典,并于1741年成为他的母校乌普萨拉大学的教授。从此以后,林奈一直从事教学和科研工作,培养了很多学生。

林奈身处的时期,正是欧洲的大航海时代,许多航海归来的生物学家和博物学家带回世界各地的动植物标本,并依据自己的喜好为之命名,从而造成一物多名或异物同名的混乱现象。并且由于各国学者语言、文字的不同,经常使植物学名十分冗长。

1753年,林奈发表《植物种志》(Species Plantarum),采用双名法,以拉丁文来为包括植物在内的生物命名。拉丁文是古罗马帝国的通用语言,不过在林奈生活的18世纪,在民间基本很少有人使用了。因为使用的人少,所以这种语言的发展变化也较少。林奈依据植物雄蕊和雌蕊的类型、大小、数量及相互排列等特征,将植物分为24纲、116目、1000多个属和10000多个种。纲(class)、目(order)、属(genus)、种(species)的分类概念是林奈的首创。他采用双名制命名法,用拉丁文来为绝大多数植物命名,即植物的常用名由两个词语组成,第一个词为"属名",第二个词为"种加词"。如果需要,还可以标注发现人或者定名人的名字首字母,以示纪念。例如:银杏学名为 Ginkgo biloba L.,Ginkgo 是属名,是名词;biloba 是种名,是形容词;最后一个字母,则是定名者姓氏的缩写,L 为林奈(Linne)的缩写。结合命名,林奈规定学名必须简化,以12个字为限,这就使资料清楚,便于整理。后来,他也用同样的方法为动物命名,这种命名法一直延用至今。

1778年1月10日,这位重新排列了动植物学界自然秩序的伟大瑞典生物学家林奈逝世。林奈的藏书和采集的标本被他的家人全部卖给了英国人,英国政府于1788年在伦敦建立了林奈学会,他的手稿和动植物标本都保存在学会中。直到今日,林奈学会仍是享誉世界的生物学会。瑞典政府为了纪念这位杰出的科学家,先后建立了林奈博物馆、林奈植物园等,并于1917年成立了瑞典林奈学会。世界顶级学府美国芝加哥大学还设有林奈的全身雕像,世界各地也都在以不同的方式缅怀和纪念这位科学家。

当我们驻足在国家自然博物馆中,请抬眼望一望展厅中悬挂的巨幅林奈画像,他坚毅的嘴唇,仿佛在向我们讲述着自然界万物的序列……

中国植物分类学奠基人——胡先骕

(1894年4月20日—1968年7月16日)

在国家自然博物馆的馆藏精品展里,有一件古老的虞美人植物标本。虞美人原产自欧洲,是著名的观赏植物。这件珍贵的标本是我国植物分类学奠基人胡先骕院士于1920年在江苏南京采集、制作并鉴定的。当时,社会治安混乱,胡先骕等人辗转浙江、江西、福建等地,相继采集到大量植物标本,他也是在江西境内进行植物科考的第一人。胡先骕以这次采集为基础,先后发表了《浙江植物名录》《江西菌类采集杂记》等一系列论文。而这件标本作为历史的见证,由当时的北平静生生物调查所植物标本室收藏,后划拨至我馆。标本历经100多年,依然保存完好。

胡先骕是江西南昌人,幼年十分聪慧,被誉为"神童"。这位小神童15岁就考入京师大学堂预科,19岁公派赴美留学。期间两度留美,先取得加州大学学士学位,后取得哈佛大学植物分类学博士学位。

第一次留美归国后,他先后任教于南京高等师范学院、东南大学、北京大学、清华大学等高校。胡先骕基于国内当时的状况和在美国所受的教育,与钱崇澍、邹秉文合编了中文教科书《高等植物学》,这也是我国第一部大学生物学教科书。

1925年,胡先骕又创办了中国最早的生物学学术刊物《中国科学社生物研究所丛刊》,用以加强学术交流。

1928年,胡先骕与秉志等人,联合创办中国科学社生物研究所、北平静生生物调查所,并担任植物部主任。当时,中国生物学才刚刚起步。北平静生生物调查所见证了我国生物学科历史从无到有的过程。

胡先骕重视发展中国的植物分类学事业,并在教育上倡导"科学救国、学以致用,独立创建、不仰外人"的教育思想。他注重培养人才,组织野外考察,收集标本和促进国际交流。中国植物志的编写也是由他最早提出的。

1933年,胡先骕创建了中国植物学会,并在庐山创建了中国第一座植物园——庐山植物园。他还是中国科学院植物研究所标本馆的创始人。

1946年,胡先骕收到朋友寄来的一些奇异的植物枝叶、球花和幼球果的标本,他根据标本反复研究并核查文献,鉴定并为这一植物命名为水杉。在此之前,植物学界普遍

生命之绿 The green of life

认为水杉是一种早已灭绝的物种,因此,水杉的发现和命名,引起全世界植物学家的震惊。1948年,水杉在庐山植物园引种成功,并被大面积种植。这种和大熊猫一样珍稀的植物,先后被引种到世界各地,胡先骕也被称为"现代水杉之父"。

毛主席曾称赞胡先骕是"中国生物学界的老祖宗"。1962年,胡先骕创作了古体科学长诗《水杉歌》,并发表在《人民日报》,陈毅副总理也盛赞该诗,并写下读后感。

胡先骕提出并发表了由中国植物分类学家首次创立的"被子植物分类的一个多元系统"和被子植物亲缘关系系统图。他的一生,出版了20多部专著,发表植物学论文140多篇,发现了植物1个新科、6个新属和100多个新种。

如今,庐山植物园的水杉高大葱郁,与水杉一起被后人铭记的还有植物学先驱胡先骕院士。

他在动荡的年代,坚守初心,在荒芜中推动中国植物学的创立,时至今日,中国植物学发展已如水杉般参天耸立,枝繁叶茂。

第 2 章 走近绿色生命的科学家
Chapter 2 Getting Closer to Green Life Scientists

中国植物的"活词典"——吴征镒

(1916 年 6 月 13 日—2013 年 6 月 20 日)

在中科院昆明植物研究所,有一块刻着"原本山川,极命草木"的石碑。1938 年,北平静生生物调查所所长胡先骕和当时云南省教育厅厅长龚自知,商定成立云南农林植物所(中科院昆明植物研究所前身),提出用"原本山川,极命草木"这 8 个字作为所训,并由龚自知书写刻石,镶于研究所墙上,后不幸被毁。

"原本山川,极命草木"出自西汉著名辞赋家枚乘的《七发》,意思是:将山川草木的名称归纳整理,使之成为文章并流传于世。这是昆明植物研究所的奠基铭,也是中国植物学工作者的理念和精神。吴征镒院士在耄耋之年重写这 8 个字,并刻石于上。"原本山川,极命草木"也是吴征镒一生与植物结缘、三下云南最真实的写照。

吴征镒祖籍安徽歙县,1916 年出生于江西九江,出生一年后,吴家迁往江苏扬州。幼年的吴征镒就如同少年鲁迅与百草园结下的不解之缘,他也在自家的后花园芜园中,观察寒来暑去与花开花落,并在观察中萌发了对植物学的兴趣。少年时期,他最爱的读物是清代吴其濬所著《植物名实图考》和日本近现代植物学家牧野富太郎所著的《日本植物图鉴》。

中学毕业后,成绩优异的吴征镒决定报考清华大学生物系,遭到了父亲的反对。当时正值战乱,有志青年们大都选择留学或报考国内顶尖大学理科专业。父亲认为花花草草无用。年轻的吴征镒却坚定地说:"我爱花草、做标本,只想学好它。"

1937 年,吴征镒从清华大学生物系毕业并留校任教。同年,抗日战争爆发,清华、北大和南开三所大学在长沙组成临时大学,1937 年底,临时大学迁往云南昆明,成立"西南联大"。吴征镒和闻一多、李继侗等老师率领学生们,历时 68 天,行程 1663.8 千米,经历诸多困苦,徒步从长沙到昆明。

在战火纷飞的年代,他积极投身抗日运动。1946 年,吴征镒加入中国共产党。在艰苦的环境中,他坚持收集植物学书籍,跋山涉水调查采集,整理书写植物卡片近三万张,每张卡片都详细记录了各种植物的学名、分布地点和相关文献。这些卡片成了吴征镒后来编纂《中国植物志》的基础资料。

他在简陋的用"洋油箱"堆成的标本室内,将没上标本台纸的许多标本,进行了系统整理和鉴定,这算是中国人自己鉴定植物标本的源头之作。他自写、自画、自印,考证完成了《滇南本草图谱》,这也是中国植物考据学的起源之作。

抗战胜利后,吴征镒先后担任中国科学院党支部首任书记、中国科学院植物研究所

生命之绿 The green of life

研究员兼副所长。1955年，他当选为中国第一批学部委员（中科院院士）。可是，心系植物的吴征镒却开始思索人生的科考之路。他怀念云南的热带季雨林、热带山顶常绿阔叶林、高山草甸、亚高山针叶林和各式各样的次生植被。他想起自己年轻时曾许下"一定立足云南，放眼中国和世界植物"的宏图大愿。

1958年盛夏，42岁的吴征镒携夫人毅然带着一双幼年的儿女迁居到云南，筹建中科院昆明植物所，开始了他与云南近一个甲子的情缘。从云南开始，吴征镒的足迹踏遍祖国的高山峻岭，踏遍非洲以外的四大洲。哪里有植物，哪里就留下过他探索的身影。

吴征镒是我国植物学家中，发现和命名植物最多的一位科学家，他改变了中国植物主要由外国学者命名的历史。由他定名和参与定名的植物类群多达1766个，涵盖94科334属，其中新属22个。以吴征镒名字命名的植物有3种，其中两种为分布在云南山区的中国特有植物，分别是"征镒冬青"和"征镒卫矛"；一种发现于湖北神农架地区，定名"征镒麻"。

吴征镒从事植物研究和教学70余年，对中国植物的来龙去脉作出了创造性贡献，他率先提出了中国植物区系的三大历史来源与15种地理成分。同学们是否知道，如果把某一地区的植物搞清楚，就要知道该地区植物的起源、形成、演化和发展。要知道中国植物在形成和演化过程有哪些类型、特点和分布，就是在追索中国植物的来龙去脉。

吴征镒主持编纂了《中国植物志》。这部长达80卷（126分册）的巨著，其中2/3卷册，是他担任主编后最终完成。此外，他还主编完成《西藏植物志》《云南植物志》等书。古人云"千里之行，始于足下"，编纂植物工具书，为后代研究者考据植物提供便捷，这是繁复浩大的工程，书中的3万多种植物，大多生长于茫茫群山中，一花一草，都是吴征镒实地考察后才收录的珍贵植物资料。

吴征镒还为我国资源合理利用与保护作出了巨大贡献。早在1958年，他就向中共云南省委省政府提出了建立24个自然保护区规划和方案。他多次向各地政府建议"营造花果山，让荒山变经济林"等相关提案，均得到采纳。他对植物资源利用和保护提出了许多卓有成效的具体意见。他向国家建议成立野生生物种质资源库，已作为大科学工程在实施进行。

吴征镒的一生，无论身处何时何地，困难都阻挡不了他潜心研究的步伐。他深知，只有自己做出成绩，才能发挥这门学科对国家的最大价值。特殊时期，他曾凭着惊人的记忆力硬生生写出了50多万字的《新华本草纲要》，其中收录了6000多种药用植物，正确率高达95%。他是中国植物的"活词典"。

吴征镒的足迹行遍大江南北，他书写了被植物界奉为典籍的诸多著述，他被中国乃至世界认为是难以逾越的植物学界科研巅峰。2011年12月10日，国际小行星中心将第175718号小行星永久命名为"吴征镒星"。从宇宙到繁星，在浩瀚的天空中，他继续揭秘着自然与植物的起源。

"杂交水稻之父"——袁隆平

（1930 年 9 月 07 日—2021 年 5 月 22 日）

我有着一个梦
埋在泥土中深信它不同
光给了它希望
雨给了它滋养
它陪种子成长

我有着一个梦
走在田埂上
它同我一般高
我拉着我最亲爱的朋友
坐在稻穗下乘凉

妈妈我来看您了
你看这晚霞洒满小山村
妈妈我陪您说说话
这种子是您亲手种下
在我心里发芽

风吹起稻浪
稻芒划过手掌
稻草在场上堆成垛
谷子迎着阳光哗啵作响
水田泛出一片橙黄

每当我们唱起这首歌，就会想到中国的"杂交水稻之父"袁隆平。《我有一个梦》不仅表达袁隆平的"禾下乘凉梦"，也表达了他对母亲深厚的感情。2021 年 5 月 22 日，袁隆平永远离开了我们，他曾经说过，"人就像一粒种子，要做好一粒种子"。江山思国土，人去

生命之绿　The green of life

稻田丰。袁隆平改变了我们的世界,又到另一个世界去实现他的"水稻比人高,禾下可乘凉"的梦想去了。

他的一生,都奉献给了祖国,奉献给了水稻,他不断创下一个又一个产量记录。我们怀念袁隆平的方式之一,就是珍惜一粥一饭来之不易,不浪费每一粒粮食。

关于杂交水稻,有许多同学们不熟悉的知识。杂交水稻到底是不是转基因作物?杂交和转基因有什么区别呢?

杂交是指两种亲本的结合,指雌雄两种水稻进行交配育种的过程。所以杂交水稻并不是转基因作物,而是选用两个在遗传上有一定差异,同时它们的优良性状又能互补的水稻品种进行杂交,有的进行多次杂交或与亲本回交,再生产具有杂种优势的杂交种,用于生产。

转基因作物是利用现代生物技术,将人们期望的目标基因,经过人工分离、重组后,导入并整合到生物体的基因组中,从而改善生物原有的性状或者赋予生物新的优良性状。

对于杂交水稻来说,被人为培养保存下来的,就是育种;如果杂交品种有哪些方面不足,就会被淘汰,再重新选育其他品种。而转基因大豆虽然长了和大豆一样的外表,但实际上已经不完全是原来的"大豆"了。

杂交水稻的基本思想和技术,以及首次成功实现是由美国人 Henry Beachell 在1963 年于印度尼西亚完成的。因此,Henry Beachell 也被学术界称为杂交水稻之父,并由此获得 1996 年的世界粮食奖。但是,由于 Henry Beachell 的设想和方案存在着某些缺陷,无法进行大规模的推广。

后来,日本学者提出了三系选育法来培育杂交水稻,以合适的野生雄性不育株作为培育杂交水稻的基础。虽然经过多年的努力,但培育的效果不是很好。在中国,60%以上的人以稻米为主食。而南方更是有 80%以上的人以稻米为主食。水稻产量的高低,对于保证粮食供应和粮食安全起到决定性的作用。

在新中国成立初期,全国水稻的平均亩产不到 150 千克,当时的粮食非常珍贵。杂交水稻在 1973 年研究成功,1976 年开始大面积推广,粮食的问题逐步得到解决。现在,杂交水稻的全国平均亩产量将近 550 千克,最高亩产量达到 1149 千克。每年因种植杂交水稻增产的粮食,可以多养活 8000 万人口。袁隆平用我国不到 9%的世界耕地面积,养活了世界上 20%的人口,创造了世界粮食史上的奇迹。

袁隆平的梦想,是用一粒种子改变世界,他一生都为了这个梦想而努力钻研。每一次科学试验,都要经过漫长的等待,半年、一年、两年……不能加速,没有捷径,日复一日,年复一年。他就像一粒种子一样,扎根泥土,不惧风雨,靠着孜孜不倦的研究和脚踏实地的努力,最终实现了这个伟大的梦想。因此,他也被誉为中国的"杂交水稻之父"。

40年来,杂交水稻技术的传授范围不断扩展,在许多国家"生根开花"。40多个国家种植了中国的杂交水稻。非洲的马达加斯加种植杂交水稻的产量比当地品种高出1倍以上,为了纪念这一伟大的成就,这个国家在新版货币印制了中国杂交水稻图案。

中国有约15亿亩盐碱地,约占世界盐碱地总面积的十分之一。其中,约有2亿亩具备种植海水稻的条件。袁隆平带领的科研团队,研究最大化利用盐碱地种植海水稻增产。这一试验成功后,海水稻亩产可达到600多千克。假如按照最低产量300千克来计算,年产粮食约300亿千克。300亿千克是什么概念呢?相当于湖南省的全年粮食总产量。

阳光炽热,土地温暖,盐碱地变良田,从此不是梦!

生命之绿 The green of life

一株小草的力量——屠呦呦

（1930年12月30日至今）

2015年10月5日，诺贝尔医学奖颁发给了中国中医科学院的屠呦呦研究员，诺贝尔科学奖项是中国医学界迄今为止获得的最高奖项，也是中医药成果获得的最高奖项。屠呦呦和她的团队在疟疾治疗研究中取得了杰出成就，这是我国第一位登顶自然科学领域奖项的本土科学家。

1930年，屠呦呦出生于浙江宁波。父亲为她取名"呦呦"，典出《诗经·小雅》篇，呦呦是鹿鸣的意思。"呦呦鹿鸣，食野之蒿。我有嘉宾，德音孔昭。"屠呦呦的名字和青蒿素，仿佛在两千年前就以某种奇特的方式连接在一起。

16岁时，她曾因感染肺结核而被迫终止学业两年，从那时起，正在读高中的屠呦呦便对医药学产生了浓厚的兴趣。1951年，屠呦呦考入北京大学药学系。在专业课程中，她尤其对植物化学、本草学和植物分类学有着极大的兴趣，这个"冷门"专业，让她最终与中医药结缘，与抗疟结缘。

1956年，全国掀起防治血吸虫病的高潮，大学毕业后在中国中医科学院中药研究所工作的屠呦呦，对有效药物半边莲（*Lobelia chinensis* Lour.）进行了生药学研究；后来，她又完成了品种比较复杂的中药银柴胡（*Stellaria dichotoma* var.*lanceolata* Bunge.）的生药学研究。这两项成果被相继收入《中药志》中。

1969年1月开始，屠呦呦领导课题组从系统收集整理的历代医籍、本草、民间方药入手，在收集2000余方药基础上，编写了640种药物为主的《抗疟单验方集》，并对其中200多种中药开展实验研究。历经380多次失败，她坚持利用现代医学方法进行分析研究，不断改进提取方法。在极为艰苦的科研条件下，屠呦呦等三名科研人员以身试药，终于发现并提取了青蒿素，开创了疟疾治疗的新方法，使全球数亿人受益，挽救了几百万人的生命。

但是，同学们是否知道？屠呦呦用于提取青蒿素的原料植物，其实并不是真青蒿，而是黄花蒿。

屠呦呦早在1987年就阐述了其中原委。黄花蒿和青蒿是两种不同的中药名称，汉代《神农本草经》中就记载了青蒿，由于古代传统中医学缺乏近代植物分类学思想与研究方法，到明代李时珍所著的《本草纲目》中，青蒿和黄花蒿出现了混乱的局面。

到了 18 世纪，日本本草学家张冠李戴，将青蒿这一名称的拉丁文名送给了另一种与青蒿素原植物互为姐妹株的蒿属植物。由于中国近代从日本翻译引进了许多植物学著作，部分学者受到日本植物学家的影响，也承认了这个张冠李戴的鉴定结果。青蒿素原植物不叫青蒿的混乱，就此传承了下来。

在编撰权威巨著《中国植物志》的时候，研究者考虑到植物学界误用青蒿之名已成习惯，所以就没有更改，仍然沿用了日本学者的错误，把青蒿一名套在了与中国本草书所描述的真青蒿毫不相干的植物上，而把李时珍时期视为一个不同于真青蒿的植物黄花蒿，送给了青蒿素原植物。

总之，中药所用的青蒿其实是黄花蒿。黄花蒿含有青蒿素，可以抗疟疾，是民间利用久远的传统药材，而植物学所称的青蒿虽然是黄花蒿的近亲，却不具备青蒿素，没有抗疟疾的功效。

青蒿和黄花蒿主要有以下区别：

青蒿常生于低海拔、湿润的河岸边等，植株高 1~2 米；茎下部叶多为 2 至 3 回栉齿状羽状深裂，中部叶多为 2 回栉齿状羽状深裂；花序呈半球形。

黄花蒿适应性强，分布几乎遍及全国，植株高约 2 米，茎下部叶多为 3 至 4 回羽状深裂，中部叶多为 2 至 3 回羽状深裂；花序近似椭圆球体。

虽然屠呦呦并非植物学家，但是仍然要把这位杰出的科学家介绍给同学们。她是新中国培养的第一代药学家，她的研发对全人类的生命健康都产生了深远的影响。屠呦呦说，青蒿素是传统中医药送给世界人民的礼物。

植物中草药为科研人员打开了一扇通向世界的大门，几十年来，中国的科学家们从未停下探索的脚步……

生命之绿 *The green of life*

身边的植物学家——陈绍煜

(1923 年 12 月 7 日—2016 年 7 月 19 日)

今天,我们漫步在色彩缤纷的植物展厅,看到与国家自然博物馆"同龄"的巨大褐藻标本,不仅感叹大自然的神奇和岁月的变迁,也应该缅怀为了这些标本呕心沥血的植物学家、原北京自然博物馆的植物室主任——陈绍煜。

博物馆的每一件展品背后,都离不开科学家们的默默研究与奉献。陈绍煜于 1923 年 12 月出生于安徽省歙县的一个徽商家庭。他七岁入私塾学习,后来进入歙县县立第一小学;1935 年 6 月以年级第二名的成绩毕业,考入徽州中学。陈绍煜直到老年,还能够背诵少年时语文老师鲍幼文的诗作"三十年前好用功,男儿何者为英雄。世间有事皆当作,天下无坚不可攻。"数学老师冯志远使用英文教科书,并以英语教授数学课,为陈绍煜的英文水平打下扎实的基础。

抗日战争期间,陈绍煜历经重重波折,于 1946 年在上万名考生中以第五名的成绩被复旦大学生物系录取。

1950 年,陈绍煜大学毕业后,在文化部科学普及局任实习编辑,1951 年,他被调入中央自然博物馆筹备处参加筹建工作。在中央自然博物馆筹备处工作期间,陈绍煜先后参与了"中国自然资源和矿产资源展""农产资源展""台湾展览"等设计提纲和展览的制作。这些展览是自然博物馆最早期的展览,但是获得了极大成功,在观众中反响热烈,当时的多位国家领导人也先后前来参观。

1959 年 1 月,自然博物馆天桥新馆正式对外开放,馆名由中央自然博物馆筹备处改为北京自然博物馆,陈绍煜负责植物研究室工作。他带领科室人员设计植物陈列,决定在陈列中重点阐明植物界的系统演化。植物陈列的内容由高等植物器官及其发展趋势、植物界进化、植物的改造和利用三大部分组成。1959年秋,植物展览如期展出,这也是新中国的第一个植物陈列馆。

1962年,陈绍煜又负责了植物陈列二馆"中国植被馆"的设计工作,并与植物室同事在全国各地收集寻获植物标本。展览中的布景箱,均使用原生产地采集、征集来的标本制作完成。

1998年,已经古稀之年的陈绍煜又参与了自然博物馆植物陈列的第三次修改。随着学科发展,这次的重建更加注重物种多样性、遗传多样性、生态多样性的原则。当全新的植物陈列对外开放时,观众可以在热带雨林生态景观中漫步,这在当时是一种新颖的展陈手段。

陈绍煜长期担任自然博物馆植物研究室主任,还曾担任过自然博物馆自然历史研究所副所长、北京植物学会理事副秘书长、中国植物学会理事等社会职务。

几十年来,陈绍煜多次到全国各地开展植物标本采集工作,并完成几十部中英文专著与论文。在陈绍煜的带领下,植物标本从无到有,到上世纪80年代末,标本室保存植物标本36000余件。

陈绍煜生前有一次接受媒体采访,被记者问及是什么精神指引着他取得丰硕的专业成果,他回答说:"我只是一只小小的徽骆驼!"他说:"所谓'徽骆驼'是指其能吃苦耐劳,像一群骆驼在漫无边际的沙漠上,走啊!走啊!我去过大西北,爬过天山,也骑过骆驼,在沙漠里不管风吹雨打,风餐露宿,向前走啊,走啊!"而我们无论逆境困境,都在苦苦追求着的每个向上的人生,是不是都有一些"徽骆驼"的影子呢?

第 3 章　绿色生命之美

Chapter 3　The Beauty of Green Life

绿色生命之美

植物是绿色自养生物，它们通过叶绿素吸收光，利用水和二氧化碳制造出构建生命体的基础物质——有机化合物。植物用光合作用供养着万物，它们分布在地球的每一个角落，是大自然的重要组成部分。地球上约有40万种植物，它们共同构成了植物世界，专门研究植物的学科，被称为植物学。某个地区的所有植物叫做植物区系。

植物有着超强的适应能力，冰点以下的积雪，某些地衣仍能存活；水温高达80度的温泉，某些藻类也能生长。从草原到高山，从森林到海洋，从极地到赤道，只要有阳光、空气和水的地方，就有植物的身影。

植物展厅不仅仅讲了植物的演化，还讲了植物与人类的关系。1951年，中央自然博物馆筹备伊始，筹备处共有7位科学家，其中，胡先骕院士是中国植物分类学的奠基人，李继侗院士是中国植物生理学的开拓者、植物生态学与植物学的奠基人。说明植物展览的科学与普及具有重要的意义。现在，就让我们从一棵苹果树开始绿色生命之美的学习之旅吧！

生命之绿 The green of life

一棵苹果树

据说当年牛顿看到苹果落地，发现了万有引力定律。那么，同学们有没有想过，牛顿看到的那棵苹果树，是由哪几部分构成的呢？

但是，在回答苹果树的问题之前，我要很严肃地告诉同学们，牛顿发现万有引力定律可不是单纯因为看到苹果落地。这个故事在牛顿生前并没有流传，虚构的成分很大，即便是真有其事，也不过是一个伟大发现的标志而已，绝不能说牛顿是因为看到一个苹果落地才发现万有引力定律。牛顿研究万有引力定律，用了 20 多年的时间。事实上，早在牛顿之前，已有不少科学家研究引力问题。天文学家开普勒发现了行星运动定律。荷兰物理学家惠更斯也在很早之前就提出了万有引力概念。

万有引力定律的发现，离不开哥白尼、伽利略、开普勒、惠更斯、胡克等科学家的研究成果，这也是牛顿在天文学、变量数学等方面研究的基础上取得的成果。牛顿临终前曾经说过："如果说我比笛卡尔看得远一点，那是因为我站在巨人的肩上。"

我们今天能够站在国家自然博物馆的展厅中，学习各种各样的生物知识，也是因为我们站在了"巨人的肩上"，是一代又一代科学家前赴后继的研究，一代又一代博物馆人的艰辛付出，才有了今天大家热爱的自然博物馆。

现在，回到苹果树的问题。苹果树作为一株完整的绿色开花植物体，是由根、茎、叶、花、果实、种子六大器官构成的。

根：具有固着和支持作用，将植物的地上部分牢固地固着在土壤中。根能从土壤中吸收水分和养料，并将水分和矿物质营养输送到植株的其他部分，让植物安稳生存。

根

其中被子植物的根一般可分为直根系和须根系两种。

直根系在正常情况下只发育成一根主根,上面长有一些侧根。有些树木长着两种类型的根系,在疏松的土壤中发育直根,在密实的土壤中发育须根。

须根系一般有很多分支,形成一个庞大的根网结构,在土壤中蔓延。这个根网能把植物牢牢固定。

茎:可以支持植物体,把叶片举向天空接受太阳光的照射。可以吸收水分和养分,并输送水分和养分达到叶片。

在茎的内部,隐藏着一个循环系统,让养分和水分在整个植物体中传输。茎的结构展现出多样性,有高耸的乔木,弯曲的藤本植物,还有地下的根状茎。有些植物还可以通过茎来繁殖,如草莓、竹子、柳树等。

叶:进行光合作用,为植物生长提供营养,并释放氧气,通常与植物的茎直接连接,或者通过叶柄连接。

叶一般分为常绿叶和凋落性叶。常绿叶常年长在植物体上,如大多数松柏类植物等,凋落性叶会随着季节的变换凋落。

花:含有植物的生殖器官,雌蕊和雄蕊。能发育成果实和种子。

果实:由花朵的子房或花托、花序轴等发育而来,是保证种子萌发的营养库。

茎

叶

花

果实

植物的演化历程

展厅讲解词介绍了一些低等植物,它们并没有根、茎、叶、花、果实、种子这样的结构。那么,植物的演化历程究竟是什么呢?

植物是生态平衡的基础。在地球上,所有的生物都是按照一定的食物关系彼此依附和关联着。这种错综复杂的关系共同构成稳定的生态关系。生物的种类越多,整个生态关系就会越稳定。

在整个生态关系中,有一类生物充当完整食物网的基础,生态学上把这类生物称作"生产者",如果没有它们的存在,那么地球上的生态系统便会崩溃瓦解。这类敢于担当的生物,就是我们十分熟悉的植物。

植物是地球上生命的主要形式之一。目前已知最古老的化石是一种原始的原核生物——蓝藻,距今已经有30多亿年的历史。植物经历了从单细胞到多细胞、从简单到复杂的演化过程,因此,植物也是地球上出现最早的生命。

蓝藻诞生于远古海洋中,是地球上最早进行光合作用产生氧气的生物。在地球早期那段漫长的时光里,它很可能是唯一的产氧生物,在海洋中释放出需氧生物赖以生存的

氧气。蓝藻的构造极为简单，它无处不在，除了湖泊、海洋、河流、沼泽等各种自然界的水域外，陆地上到处都有蓝藻的身影，甚至连荒漠里也能找到它的踪迹。

不过蓝藻最富集的地方还是浅水环境里的微生物席。微生物席是一种由微生物、微生物分泌的胞外聚合物以及它捕捉黏结的物质共同组成的席状物，至少在 30 亿年前就已经出现了，它们组成了地球上最早的生态系统。

微生物席的组成比较复杂，在其上层进行光合产氧作用的部分里，蓝藻是主力军。我们把蓝藻归于原核生物中，蓝藻也被称作蓝细菌，它只能被称为广义的植物。海洋中的蓝细菌、原绿藻等原核生物逐渐演化出了其他藻类，现在有 3 万多种。

除了藻类外，还有一类较为低等的广义植物类群——菌类植物。我们一般说的菌类植物是指真菌家族的蕈(xùn)类，就是大家熟悉的"蘑菇"。这些生物多数不进行光合作用，而是吸收非自身的有机物来生存，在食物链中，蘑菇属于分解者，它们没有种子，通过释放孢子来进行后代繁殖，属于孢子植物。

那么，是不是只有低等植物才会用孢子进行繁殖呢？

其实并非如此，高等植物中，苔藓类和蕨类植物也是依靠孢子来繁衍的。目前全世界苔藓类有 215 科，1250 多属，2 万余种。分为苔类门，藓类门和角苔门。它们是仅次于被子植物的第二大类群，数量远远多于蕨类植物和裸子植物。它们拥有高等植物的标志性结构——胚。

苔藓类植物生长在阴凉潮湿的地方，虽然可以进行光合作用，但是没有真正的根、茎、叶等植物器官。它们也没有承担植物体支撑和运输作用的维管束。

苔藓植物是一类很有趣的生物，在繁殖期到来时，许多苔藓类植物都会长出一个个"小棒子"，这就是它们的精子与卵子融合形成的孢子体。在孢子囊体内有许许多多的孢子，待到成熟时，孢子囊就会自动打开，这些新生的孢子便会出来，开始新的生命之旅。可能有许多人讨厌这些小家伙，因为常常会踩到它们而滑倒。事实上，我们应该为见到苔藓类植物高兴，因为许多苔藓类植物可作为空气质量的检测生物。换言之，当你见到大量的苔藓类植物时，说明你所处地方的空气质量相当不错。

地钱是一种典型的苔藓类植物，它分布十分广泛，无论是公园还是小路边，都可以看到它的身影。1753 年，瑞典分类学家林奈用自己创建的生物双命名法体系，给地钱取了一个学名：*Marcbantia polymorpba* L.。这意味着它是最早一批被发表的苔类植物。因为它贴着地面，向四周呈分叉状，好像钱币的形状，所以中文名称叫地钱。

地钱的属名 Marcbantia 表明它的分类学家世，是属于地钱属，种加词则相当于它的名字，polymorpba 的含义就是形态多样，表示该物种具有形态多变的特性。正因为它的形态多变，所以才练就了高超的生存本领。

生命之绿 *The green of life*

地钱是雌雄异株,雌株会长出雌生殖托,成熟时产生一个卵细胞。雄株长出雄生殖托,产生许多精子。进入繁殖阶段后,雌株和雄株高高伸出生殖托,仿佛举着一把把小伞,这可不是为了给自己挡风遮雨,而是通过增加高度,将精子和孢子散播到更远的地方。除精卵细胞结合发育成子代的有性生殖方式外,地钱还有一种从叶状体上生出好像迷你口杯的特殊构造——胞芽杯,来进行无性繁殖。胞芽杯内生芽孢,芽孢成熟后,就可以借助风吹雨打等外力离开母体,到新的环境中萌发成为新的地钱个体。

蕨类植物是有着辉煌历史的古老植物,在植物进化史上具有里程碑式的意义。首先,蕨类植物中出现了维管束,在结构支撑、养分运输上产生了极大的优势,这也是蕨类植物之所以能迅速崛起的动力。虽然现今的蕨类植物已经失去了以往的繁盛,但我们仍然可以在很多地方见到它们的身影。

在北京环境稍好的公园内,如天坛公园、北京植物园、百望山森林公园等都广泛分布着一种蕨类植物,中文学名叫"卷柏",它们已经在地球上生活了长达4亿年,是迄今为止陆地植物中最古老的种类。在漫长的演化过程中,卷柏的外貌形态和祖先高度一致。它也被称为"九转还魂草",卷柏的内部细胞原生质能产生抵御干旱的化合物——海藻糖,帮助细胞在脱水时维持较稳定的新陈代谢。所以从热带雨林到荒漠戈壁,常常能看到卷柏属植物的足迹。其中,垫状卷柏的"还魂"能力更胜一筹。当水分充足的时候,卷柏的枝叶舒展翠绿,当水分不足时,它就失去颜色,蜷曲抱团宛如枯死一般。曾有生物学家用卷柏做成植物标本。11年后,把它浸入水中,居然又恢复了生机。

下次,当我们再见到这些低调的植物时,可以驻足好好观察。毕竟,这些生命承载地球中久远的记忆……

如何划定低等植物与高等植物

一般来说,要判断一种植物是低等还是高等,取决于它是否存在一个重要的器官——胚,而不是看它是否利用孢子繁殖。胚就是受精卵在母体内发育成熟的幼小植物体,包括胚芽、胚轴、胚根和子叶等部分。

种子植物分为裸子植物和被子植物。这里的"子"是指植物的种子。

裸子植物的胚珠外面没有子房壁包被,不形成果皮,种子是裸露的,所有的裸子植物都是木本植物。

被子植物是植物界中最高等、最繁盛的家族,种子外包裹着植物的器官,并且拥有真正意义上的花。被子植物有木本、草本之分。

区分裸子植物与被子植物有两个小窍门:

①看看种子植物是否是草本植物,若是,就是被子植物;

②若看到的种子植物为木本植物,则看是否有真正的花,若有,就是被子植物;若没有,就是裸子植物。

植物的分类比较多,一般从广义上来划分。有分成种子植物、苔藓植物、蕨类植物和菌藻类植物四大类的;也有按照植物的生存方式:分为藻类植物、菌类植物、地衣类植物、苔藓类植物、蕨类植物以及种子植物六大类的;还有按照苔藓类植物、石松类植物、蕨类植物、裸子植物和被子植物区分的。

其中,除了异养型的菌类,其他都是生态系统中的生产者,即构建食物网的基础类群。

种子植物在这六大类植物中等级最高,而种子植物中的被子植物又比裸子植物等级高。

生命之绿 The green of life

植物界的"猛兽家族"

在植物界，有一类"猛兽家族"，它们专门以捕获昆虫为生，并通过消化昆虫来获取营养，所以人们把它们称作食虫植物。其实，食虫植物的捕猎范围并不限于昆虫，还包括其他小型节肢动物和环节动物，所以，人们也叫它们"食肉植物"。

目前，被人类已知的食虫植物约有 630 种，它们大多生长在土壤贫瘠的环境中，特别是土壤缺少氮元素的地区。因为缺乏生命活动所必需的氮、磷等营养元素，它们为了适应生存环境，不得不进化成植物中的"猛兽"，长出诸如捕虫夹、捕虫笼等器官，抓捕昆虫和其他小动物"补充营养"。

近年来，这些奇特的植物走进了人们的生活，成了家庭观赏养殖的植物新宠。一方面，人们被食虫植物的奇特本领所吸引；另一方面，商家经常在夏季打出"天然捕蚊能手"的广告诱惑人们购买，因此，这些舶来品成为了我们身边的风景，如花市捕蝇草、红瓶猪笼草、查尔逊瓶子草、金丝绒茅膏菜……

其实，中国也生长着一些原生食虫植物，它们分属于瓶子草目和管状花目 2 目；猪笼草科、茅膏菜科、狸藻科 3 科；猪笼草属、茅膏菜属、貉藻属、捕虫堇属、狸藻属 5 属，共计 27 种。

中国原生的猪笼草只有一种，属瓶子草目、猪笼草科、猪笼草属，拉丁种名翻译为"奇异猪笼草"。它产于广东省、海南省和香港特别行政区，是分布比较广泛的猪笼草属物种。它的捕虫方式也和其它猪笼草一样，靠叶顶的"瓶子陷阱"。它们先通过"瓶盖"分泌的蜜汁引诱那些喜欢甜味的昆虫，昆虫被猪笼草的蜜汁吸引后，就会沿着"瓶子"的内壁爬下去吃蜜，当昆虫吃饱后准备原路返回时，光滑的内壁却让它们没法往上爬，由于"瓶子"特殊的形状和结构，"瓶底"还有个"死亡尖角"，使昆虫易进难出，昆虫越是挣扎就越容易滑落瓶内，进而被消化液淹死，最终被猪笼草含有蛋白酶的消化液消化得尸骨全无。

有人问，猪笼草的"瓶盖"会在昆虫掉进"瓶子"里时自动盖上吗？其实，"瓶盖"并不会关闭，"瓶盖"的作用只是为了给猪笼草的"瓶身"遮风挡雨，防止瓶子里的液体被稀释。

随着近年来广东、海南等地沿海湿地的开发，奇异猪笼草的分布范围和种群日益缩小，并且这种本土原生植物移栽到花盆中几乎无法靠人工养殖成活，因此，它也是国家重点保护植物。

家庭培育的观赏猪笼草通常是温室栽培的红瓶猪笼草，这是一种大自然中没有的植物，是由翼状猪笼草和葫芦猪笼草杂交出来的园艺品种。很多人在花卉市场购买猪笼草在家庭养殖,用来捕捉蚊蝇,其实,猪笼草特殊的蜜腺最容易吸引的是蚂蚁,由于缺乏特定环境,猪笼草在家中是很难捕捉蚊蝇的。

茅膏菜的根、茎、叶、花都有毒,是名副其实的"五毒俱全",它的叶子上含有对昆虫足以致命的黏液,真可谓是"植物猛兽"中的翘楚。

茅膏菜分布于我国长白山地区、长江流域、珠江流域,以及西藏南部。它的每片叶子上都长着一层浓密的腺毛,每根腺毛的顶端都有亮晶晶的黏液,昆虫一旦停在叶片上,就会被黏液黏住,然后,叶片外围的腺毛就开始卷曲,将昆虫紧紧地压在叶片上,腺毛分泌蛋白质分解酶,把昆虫逐渐消化殆尽,此后,这些腺毛又重新张开,再次分泌新的黏液,循环往复。

茅膏菜还具有"团队精神",要是俘获的昆虫个头大,茅膏菜的叶子会自动对折夹住它,假如一片叶子的力量不够,其他叶子就会过来"帮忙",不会让昆虫轻易逃脱。根据研究,茅膏菜具有类似神经传递的反应。捉到昆虫后,信号就会沿着布满叶子的叶脉方向传遍整株植物,科学家将直径仅 0.2 毫米的头发丝放在它的叶子上,叶面的腺毛也能立刻感应弯曲。

狸藻属于管状花目、狸藻科、狸藻属。全世界约有 230 种狸藻,根据较新的研究,我国分布有 26 种狸藻属植物。这类植物在命名上,水生的叫做"狸藻",而陆生或附生的种类叫"挖耳草",多数生长在长江以南,少数分布于长江以北,其中,怒江挖耳草是中国特有种。

狸藻的捕猎方式是依靠捕虫囊,捕虫囊有可以开合的瓣膜,瓣膜的外侧长着感应毛。当猎物停留在囊口时,半瘪的捕虫囊就会迅速鼓起来,形成一股吸力,同时瓣膜打开,就好像吸尘器一样,把猎物吸入囊中,整个反应过程耗时大约 0.01 秒,比起人们熟悉的同样靠运动出击的捕蝇草快至少 200 倍。

在北京海淀区的翠湖湿地公园、圆明园等地,生活着黄花狸藻,感兴趣的同学,可以到公园里寻找这种奇特的食虫植物。

这些高度特化的"植物猛兽",是大自然的一部分,然而四通八达的交通网和滚滚的车流,正在破坏着食虫植物的栖息地。保护生态环境是个宏大的概念,作为普通人的我们也许很难做些什么,但我们至少可以记住,如果在野外邂逅了这些食虫植物,除了照片和回忆,什么都不要带走。

生命之绿 The green of life

植物界的"大熊猫"

2023年8月8日,在中国成都大运会闭幕式上,一棵巨树扎根于舞台,金色根系从观众席一路生长,白鸽般的花朵在枝头摇曳。

这棵树的灵感来自于有着"植物界大熊猫"之称的珙桐树,它也是贯穿大运会闭幕式的主体视觉形象。

珙桐(*Davidia involucrata* Baill.)是蓝果树科珙桐属,落叶乔木,因为花朵盛开时,外形洁白、形似白鸽展开的翅膀。在山风拂动中,宛如满树白鸽振翅欲飞,故得名"中国鸽子树"。

鸽子寓意着和平与希望,而形似白鸽的珙桐也被认为是和平的使者。

其实,令人赏心悦目的"鸽子花",并不是珙桐真正意义上的花。而是由叶片特化而成的两片独特的、大小不一的苞片。珙桐真正的花,是暗藏在苞片下面、呈暗紫色的头状花序。夏季花序初开时,这些苞片并不是白色的,而是如同叶片一样泛着新绿;随着盛花期到来,逐渐呈现为乳白色,而后在末花期,转为棕黄色凋落。

千万不要小瞧苞片的作用,苞片不仅有洁白优雅的外形,它自然展开后的颜色和气味,可以吸引昆虫前来传粉。苞片还是花序的防护伞,能有效避免狂风暴雨和太阳辐射对花序的伤害,让昆虫有机会接触到更多花粉。

颜值与实力并存的珙桐,是距今6000万年前新生代第三纪古热带植物区系的孑遗种,是我国最具代表性的孑遗植物之一,被誉为"植物活化石"。在第四纪冰川时期,地球上大部分地区的珙桐相继灭绝,仅在今天的四川、贵州、云南、湖南以及湖北等地区幸存。它是中国特有的单属植物,是国家一级重点保护野生植物,也是全世界著名的观赏植物。

大运会闭幕式上,时任四川省省长、成都大运会组委会执行主席黄强在致辞中分享了一个历久弥新的故事——1954年,周恩来总理在瑞士日内瓦湖畔,欣喜地看到了引自中国的珙桐树。

其实,早在1869年5月,法国传教士戴维(P.A.David.)来到四川省雅安市宝兴县穆坪镇的一个山沟里。他在四处观望时,突然被一片片硕大的花瓣惊呆了,这些花瓣在绿叶中随风舞动,宛如栖息在枝头的白色鸽子。戴维被它的美丽震撼,决定采集标本,并带回法国。

这位戴维神父,作为西方人首次发现珙桐,并为它命名拉丁种名。值得一提的是,这位神父也是为麋鹿命名拉丁种名的人。

1900年,植物猎人威尔逊受英国园艺公司派遣来到中国,开启了他的中国物种探索之旅。年仅24岁的威尔逊,在当年5月来到中国西南部一片森林考察时,意外被一根横着的树杈绊倒,他发现这正是他苦苦寻找的珙桐……

1904年,威尔逊将珙桐引入欧洲和北美洲,从此,这种美丽的鸽子树在欧洲开花结果,成为著名的观赏树。

四川与珙桐有着不解之缘——珙桐最早的发现地和珙桐模式标本均产自四川省宝兴县,2008年,在四川省荥经县龙苍沟乡又意外发现了近10万亩野生珙桐群落。目前,四川全省珙桐树的数量居全国第一。

近年来,珙桐在中国逐渐被引种为观赏植物。北京植物园栽培的珙桐,是中国大陆地区,珙桐陆地栽培的最北位置。

珙桐与大熊猫一样,都是我国宝贵的物种资源。由于珙桐的种子自然萌发困难、幼苗存活率低,再加上生存环境的变化,使得它的野外天然分布面积和数量急剧下降。为保护这一古老的孑遗植物,我国于1984年把珙桐列入首批8个重点保护的一级濒危物种,并把多个分布区划为国家级自然保护区。

古老的历史赋予了珙桐特殊的生物学和生态学价值。珙桐之所以珍稀,不仅因为它经受了寒冷气候的考验,还在于它对生存环境的特殊需求。作为宝贵的物种资源,珙桐在科学研究、园林应用和文化寓意方面都具有深远意义。

这种洁白无瑕,饱含深意的植物,矗立在中国西南部的密林深处,矗立在世界的角落中。它像是穿越古老时空的友谊使者,轻轻地诉说着地球的历史沧桑。无论过去、现在还是将来,生命茁壮、和平永恒。

生命之绿 *The green of life*

恐龙的见证者

国家自然博物馆的前身是北京自然博物馆,馆徽上的动植物分别是马门溪龙和银杏叶,这是中国特有的古老的中生代古爬行动物和古植物。

国家植物标本馆的馆徽也以银杏为设计元素。

《中国植物志》共有 80 卷,每一卷的封面上都印着银杏。

银杏是地球上现存最古老的大型植物之一,在距今 3 亿多年前的石炭纪,它就遍布在地球上的每个角落。它曾经在中生代与恐龙相伴了 1 亿多年的时光,并在鼎盛之后逐渐走向衰落。大约在距今 500 多万年前,银杏在北美洲灭绝。大约在距今 260 多万年前,银杏在欧洲灭绝。

地球经历了寒冷的冰河时期,银杏的栖息地被逐渐压缩,最后,这种古老的植物仅存于中国。它承载了地球的历史变化,见证了恐龙的兴衰。又和人类一起,走进了新世纪。

目前,银杏门下只有 1 纲 1 目 1 科 1 属 1 种——银杏。它有着独一无二的扇形叶片,叶脉分布呈叉状脉序,这与其他种子植物的网状脉和平行脉十分不同。每年春夏季节,叶片的颜色青翠欲滴,而每到秋季,又变成金黄。银杏的种子叫白果,据《本草纲目》中记载,白果苦中带甜,生食有微量毒性,能治疗咳嗽、哮喘,还能通畅血管,改善大脑功能等。银杏叶提取物对治疗冠心病、心绞痛和高血脂有一定效果。

银杏是中生代孑遗植物,也是中国特有树种,被称为植物界的活化石。它不会开花结果,是裸子植物中唯一的阔叶落叶乔木。它属于雌雄异株,其性别由性染色体决定。雄株具有 XY 型染色体,雌珠具有 XX 型染色体。

银杏的树形高大优美,树干直径 4 米左右,高度可达 20~40 米,银杏也叫"公孙树",中国各地有许多百年甚至千年以上的老树。山东省日照市莒县浮来山定林寺、浙江省杭州市的天目山、贵州省福泉市黄寺镇李家湾村等地,都有树龄超过 4000 岁的银杏古树。

那么,银杏树长寿的秘密是什么呢?科研人员经过研究发现,银杏通过细胞分裂和物质合成提高树干密度和强度。并通过关键基因的持续表达和重要代谢物的日积月累,不断增强树体的抵抗力和韧性,使得树干形成层干细胞不进入衰老期,因此,千年古树才能始终保持健康活力的状态。

成都市和丹东市都把银杏作为市树。每到秋天,满树金黄绚烂,吸引了古今国人们深情的注视。无数文人墨客写诗吟诵它的身姿。其中,宋代文豪苏轼就写诗赞颂过银杏清秀如画,一树擎天。

银杏,这个恐龙时代的见证者,在这如画的祖国壮美山河里,继续见证着人类日新月异的今天。

生命之绿 The green of life

天安门的大花篮

每到国庆节,首都北京天安门,都会被缤纷绚丽的花朵装点得分外隆重。而提起天安门的花,大家就会不由自主想到广场上的大花篮!从1986年起,天安门就开始摆放立体大花篮。一个个花篮,见证了祖国的繁荣发展和时代的变迁,满载着人们对祖国的深切祝福,更是国人心底里,对于国庆特殊的记忆。

那么,大家知道每年国庆的时候,天安门花团锦簇的大花篮里面都有些什么花吗?

每年的大花篮内,花朵的种类和数量都不固定。大体上说,天安门的大花篮里,大约有30种花每年轮流摆放。这些都是我国各省的省花、市花或者区花,以及特别行政区的省花、市花或区花。

有时候,还会搭配一些水果,比如石榴象征着全国各族人们紧密团结在一起。葡萄、苹果、橘子等象征着神州大地瓜果丰收的景象。

中原地区的省花有河北省省花太平花、山东省省花牡丹、河南省省花腊梅、山西省省花榆叶梅、陕西省省花百合、湖北省省花梅花、安徽省省花黄山杜鹃、湖南省和四川省的省花木芙蓉、江西省和贵州省省花杜鹃花。

沿海地区的省花有江苏省省花茉莉花、浙江省省花兰花、福建省省花水仙、广东省

省花木棉花、海南省省花三角梅、台湾省省花白蝴蝶兰。

西北地区的省花有甘肃省和青海省省花郁金香。

东三省的省花有辽宁省省花天女花、吉林省省花君子兰、黑龙江省花丁香。

还有一些区花和市花，如宁夏回族自治区和内蒙古自治区的区花马兰花、新疆维吾尔自治区的区花天山雪莲、西藏自治区的区花报春花、广西壮族自治区的区花桂花、香港特区的区花紫荆花、澳门特区的区花莲花。

北京市的双市花菊花和月季、上海市的市花白玉兰、天津市的市花月季、重庆市的市花山茶花、沈阳市和拉萨市的市花玫瑰花等等。

这些花用不同的特性和不同的文化内涵，代表了不同的城市文化。

希望下次国庆的时候，大家再去天安门广场的时候，能在美丽的大花篮里一一对应，找到这些美丽的花。

生命之绿 The green of life

中国标本馆的前世今生

1753年，瑞典生物分类学家林奈（Carl Linnaeus）编写出版了《植物种志》，在书中，他根据植物雄蕊等特征，将约7700种植物划分为24个纲，建立了一个当时受到广泛应用的植物分类系统。

16世纪，意大利波罗尼亚大学成立了世界上第一个植物标本馆。目前，根据世界标本馆索引记载，全球176个国家有2962家标本馆，共收藏了近3.8亿份标本。

其中，全球规模最大的标本馆是法国国立自然历史博物馆，馆藏800多万份标本，它建立于1635年。主要收藏全球性标本，标本来自非洲、欧洲、马达加斯加、东南亚等地，并以保存法国传教士19世纪采自中国西南各省区的标本著称。

世界上最小的公共标本馆在马达加斯加北部的一个岛国上，它是塞舌尔标本馆，仅存500份标本。

中国被称为世界园林之母（China.Mother of Gardens）。在过去的两百多年中，外国人以传教和考察的名义，在中国采集了近100万份标本，其中许多是珍稀的植物模式标本。

目前，这些标本主要集中在英国大英博物馆、邱园植物园标本馆、爱丁堡植物园标本馆、法国巴黎自然历史博物馆、德国柏林植物园标本馆、俄罗斯圣彼得堡植物园标本馆等几大标本馆中。这些来自中国的标本，对他们本国的标本馆和植物分类学的发展贡献十分巨大。

而这一切，都与中国无关。

随着近代运输行业的发展，许多商品从广东沿海运出去，因为亚洲的香料很多，这个输送香料的港口就叫做香港。

1841年，英国医生海因兹随船来到中国香港，采集了140多种标本，此后又有许多外国植物学家到香港进行标本采集，只可惜这些标本都被带离中国香港。直到1878年，香港才成立了植物标本室。

目前，中国已建立300多家植物标本馆。有些标本历经百年，有些标本刚刚采集。每一次，我们走进标本馆，就如同翻开一部植物王国的百科全书。形形色色琳琅满目的标本，好似书中的文字，记录了地球历史中植物的演化历程。

在我国，超过100万份标本的大型标本馆有三家，分别是中国科学院植物研究所植物标本馆，馆藏标本280余万份；中国科学院昆明植物研究所植物标本馆，馆藏标本150余万份；中国科学院华南植物园标本馆，馆藏标本100余万份。

标本馆是价值连城的科学宝库，它储存着植物的一切身份信息，也是进行植物标本收集、保存和编目的场所，是植物学家及植物爱好者学习和研究的机构。

从1959年开始，以吴征镒为首的200多位植物学家就开始编著《中国植物志》，到2004年，浩瀚的巨著完成。这部80卷的图书收录了3万多种植物，它就是根据我国这些植物标本馆收藏的数百万件植物标本编写而成。

《中国植物志》就像植物的户口本，它是社会上其他学科利用植物资源的工具书，也是植物的大词典。通过查找植物志，我们就可以了解到相关的植物信息。

当一种植物第一次被发现或命名为一个独立的物种，分类学家把它描述发表，在给植物起名称时，要依据一份标本，这份标本就被指定为模式标本。普通的标本我们可以到野外再次采集，但是模式标本就是一份永远保存不可再生的，如果它消失了，很多信息就消失了。

过去我们依靠标本的形态特征来鉴定植物，可能会出现错误，有些物种的存在是以种群的形式，种群分布在不同的生态环境，叶片的大小、颜色和形状等很多性状就会因为环境而发生改变，而植物进入分子时代，植物的DNA分子序列相对稳定，能够准确提供植物的遗传信息。

这也是科技带给我们的收获。

现在出现了数字图书馆，人们只要接入互联网就可以获取信息。数字植物标本馆又叫虚拟植物标本馆，它是植物标本馆的数字化形式。中国数字植物标本馆简称CVH，这是一个跨部门的共建网站，成员单位已达76家，包括中国科学院和地方科学院及一些大专院校标本馆，基本上包含了我国主要和重要的标本馆。目前，数字标本已经达到670万份。这说明现代植物学已经进入分子时代。

但是，标本的意义永远不会改变。

生命之绿 *The green of life*

标本制作——化绿色生命之美为永恒

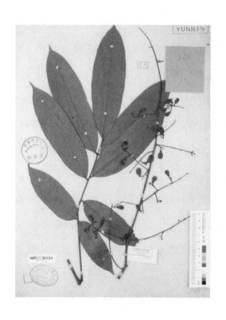
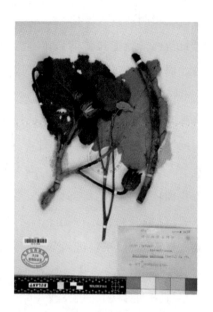

在国家自然博物馆收藏的37万件标本中,植物标本有8万余件。在这些标本中,有许多珍贵的孤品标本。如古黄河象头骨化石、真骨率达90%以上的体长26米的巨型井研马门溪龙化石、前越南领导人胡志明送给毛泽东主席的亚洲象、新西兰坎特伯雷国家博物馆赠送给我国的恐鸟骨骼标本、2009年自然博物馆古生物学家发现的震惊世界的中华侏罗兽化石、历经一百余年的虞美人植物标本、朴树标本等等。

展厅中展示的标本,只是博物馆众多馆藏中的凤毛麟角。透过化石的印痕,我们能看到那一个个鲜活的远古生命,似乎穿越了时空,聆听着远古地球上遥远的声音。植物陈列厅则像一部绿色的史诗,叙述着植物亿万年的演变。植物由水生到登陆,一朵花的盛开,一粒种子的传播,都蕴藏了无数的奥秘,留给我们无数的想象和疑问。

植物展厅中长达4米的海黍子和2米长的海带标本,是1959年中国科学院海洋研究所为了献礼国庆十周年,无偿捐赠给自然博物馆的。经历了半个世纪的风雨,藻体已经有了裂痕,但岁月留下的痕迹并没有削弱标本的魅力,反而增加了它的内涵和时代感。这些具有特殊意义的标本,在向中国的第一个系统植物学展览致敬。那么,展厅中的标本,究竟是如何制作的呢?

国家自然博物馆的馆藏标本,按类别大致可分为 6 组:植物组标本、昆虫组标本、化石组标本、鸟类假剥制标本、液浸标本和兽类标本。

植物作为自然界中庞大的类群,标本的采集与制作就显得格外重要。最常见的植物标本是腊叶标本,"腊"的意思就是"干",腊叶标本也叫压制标本。腊叶标本对于植物分类意义重大,它使得植物学家们一年四季都可以查对采集自不同地区的标本,并借助这些标本对植物进行描述和鉴定。16 世纪后期,植物分类学的迅速发展,在相当大的程度上是由腊叶标本促成的。

在国家自然博物馆的植物展厅中,就有一面"叶片墙",上面展示了 127 种被子植物的 137 件不同形态的腊叶标本。对于我们普通人来说,在生活中把自种的花草制成标本或者书签,是对生命的热爱和对美丽大自然的留恋。

现在,我们就来介绍一下植物腊叶标本的制作方法。

植物腊叶标本制作也称为压制标本,一般都是将新鲜植物材料用吸水纸压制,干燥后装订在白色硬纸上,我们把这种白色硬纸称为台纸。完整的腊叶标本制作分为采集、整理、压制、上台纸等步骤。

采集时,需要携带标本夹、枝剪、铁铲、记录本、采集号签等装备。我们要注意审美、布局、结构和叶片的平整性,这样才能把它压制得很漂亮。

我们应该认真填写采集记录单,记录采集时植物的状态,应特别描述一下植物在压制后失去的一些特征,如植物器官表面是光滑还是有毛有刺,植物的气味、颜色等。采集时,要从数株同一种类的植物中选择各器官最完整的植株做标本,要优先采集有花和果实的部分,并体现其分支。先剪去残叶,再把过于浓密的部分有选择地适当剪掉,这样不

表1、植物采集记录单:

采集号			
采集时间			
采集人			
采集地点			
海拔高度			
树高		胸径	
树皮		树枝	
叶		花	
果		习性	
生态环境		用途	
中文名		俗名	
学名		科名	

表2、标本标签:

编号	
中文名	
科名	
学名	
采集地点	
制作人	
制作日期	

生命之绿 The green of life

至于堆积,也能使一个立体的实物变成平面。

一般情况下,所采标本大小以 25~30 厘米为宜。

高大的草本植物可将其折成 V/N 或 W 字形压入标本夹。

如遇到巨大叶片的植物,如牛蒡等,在一张纸上压不下的情况下,可以将一片叶子分成两份。

雌雄异株的植物,要分别采集雌株和雄株。分别编号并注明它们之间的关系。雌雄同株的植物,最好两种花都采集。

同株植物中不同形态的叶子,要分别采集。

寄生植物尽量把寄主一起采下,如菟丝子,桑寄生等。应当连同寄主一起,尽量不要将两者分开。

有地下根、块根的植物,如百合科,天南星科等,要把地下部分挖出来编号保存。

除非特殊需求,一般不采集带有病菌或受到机械损伤的枝条作为标本。

腊叶标本应能完整展示植物的形态特征。

种子标本必须有花和果实。

蕨类植物标本叶上必须有孢子囊。

苔藓植物标本必须有叶片和孢子囊。

采集的时间和地点务必记录清晰,这样有助于区别各种植物的生活习性、生态环境、分布规律等。还有一点很重要,一般在同一个地点、时间采集的同一物种标本,最好采集 2~3 份,使用同一个编号,这样有助于我们更好地保存和观察。

所以,每份标本应该有编号和野外记录,内容包括名称,产地生态环境,海拔高度,植株高度,叶、花、果的异变性状,如颜色香味等。此外还应记录分布数量和经济用途,采集人姓名和采集年月日等信息,将采集记录粘贴在台纸的左上角或右下角位置,这样才能为今后的研究和观察提供可靠的原始资料。

压制植物标本是关键步骤。刚采集来的标本比较湿,我们需要用吸水纸把植物材料压干,越快越好,如果没有吸水纸,用报纸压干也可以。先把装有植物标本的标本夹放在平整的桌面上,展平标本,在标本上放上 5~8 层吸水纸,再放植物标本,然后再覆盖 5 层左右的吸水纸,再放下一份植物标本,依次循环,直到摆放好最后一份植物标本,并在最上层覆盖 5 层左右的吸水纸。摆放时,尽量保持平整,避免因为植物标本摆放导致中间高,四周低的情况,一旦植物标本接触到空气,就容易变黄或者发皱。

最后,我们把另一扇标本夹放在最上面,压紧并用绑带固定牢固。为了尽快去除植物标本中的水分,可每天更换吸水纸,大约 10 天左右,干制标本就完成了。

春夏季节的植物含水量高,干制的时间会长些,秋冬季节的植物含水量低,干制的时间可能短些。我们在制作过程中,可以根据标本情况具体斟酌。为了防止植物标本干制后被虫蛀,可以将其放入恒温干燥箱中烘干 1~2 天,如果没有恒温干燥箱,也可以在制作完成的植物标本中放入干燥剂和驱虫剂。

最后,我们把干制的标本固定在白板纸上,根据标本尺寸,选择不同尺寸的白板纸,摆放时,尽量使干制标本和台纸边缘保留 2cm 以上的距离。一般情况下,标本摆放应向右倾斜一定角度,并在右下角或左上角贴标本标签和采集记录单。

如果想制作植物书签,干燥后塑封保存就可以;如果只是为了观赏,买一些滴胶材料,我们也可以自制滴胶的鲜花植物标本,化瞬间为永恒,留住美好的大自然。

但是,一定要记得,珍稀植物千万不要随便采集!

第 4 章　中国文化中的植物

Chapter 4　Plants in Chinese Culture

第 4 章 中国文化中的植物
Part 4 Plants in Chinese Culture

中国文化源远流长，最古老的诗歌总集《诗经》中，共有305篇诗作。因为《诗经》是我国古代北方的文学作品，描述区域以黄河流域为主，所以各篇章中出现的植物主要分布于北方。

这些篇章中，有153篇提到植物或者描述植物。比如，"桑"是《诗经》中出现篇目最多的植物，有20篇；"黍"是古代的五谷之一，《诗经》中有17篇提到它；有12篇提到了"枣"；9篇提到了"小麦"；7篇提到了"松""葫芦瓜""葛藤""芦苇""大豆""柞木"等物；6篇提到了"大麻""稻""粟""枸杞""黄荆"等物。

《诗经》中还有我们熟悉的许多植物相关日常语，如"投桃报李""桃之夭夭""参差荇菜"等。

《楚辞》中描绘的植物达到101种。其中香草就有23种，比如"白芷""泽兰""熏草"等。特别是屈原的作品，常常托物言志，用香草比喻君子，用"蒺藜"比喻小人。

魏晋南北朝盛行"骈体文"，最著名的文章就是陶渊明的植物文章《桃花源记》："晋太元中，武陵人捕鱼为业。缘溪行，忘路之远近。忽逢桃花林，夹岸数百步，中无杂树，芳草鲜美，落英缤纷，渔人甚异之。复前行，欲穷其林。"

汉代乐府诗《孔雀东南飞》中写道："君当作磐石，妾当作蒲苇。蒲苇韧如丝，磐石无转移。""蒲苇"是指两种植物，蒲是蒲草，苇是芦苇。用两种植物的特性，来比喻坚贞的爱情。

到了唐宋时期，诗歌和词赋达到鼎盛，其中表现植物意向的内容接近诗词总数的一半。全唐诗共5万余首，描写植物398种。其中，柳树是引用最多的植物，占3463首。王维的《渭城曲》就形象的描述了柳树："渭城朝雨浥轻尘，客舍青青柳色新。劝君更进一杯酒，西出阳关无故人。"

其次，竹和松这两种植物也被引用达三千多次。宋代诗词中，梅花大量出现，王安石著名的《梅花》："墙角数枝梅，凌寒独自开。遥知不是雪，为有暗香来。"

元曲中马致远的《天净沙》："枯藤老树昏鸦，小桥流水人家，古道西风瘦马。夕阳西下，断肠人在天涯。"枯藤与老树构成了萧瑟的场景。

明清小说中，各种植物意向层出不穷，《红楼梦》的大观园中不仅有各种花草树木，小说中还以花喻人，每位红楼女儿都有花语判词和谶语。比如，晴为黛影，晴雯和黛玉都是芙蓉花。宝钗是牡丹花，湘云是海棠花，袭人是桃花等等。书中又根据每个人物性格特点，让她们吟咏作诗，比如结下海棠诗社后，大家做海棠诗、梨花诗等。小说中著名的《葬花吟》和贾宝玉祭奠晴雯的《芙蓉女儿诔》都是吟诵植物的名篇。

历史上，还有许多关于植物的典故。《晏子春秋》有一个故事，叫"二桃杀三士"。讲的

生命之绿 *The green of life*

是春秋时,公孙接、田开疆、古治子三个人,都是齐景公的大臣,都以勇猛闻名,但是这三个人平时我行我素,文化素质不高,还互相妒忌争抢功劳。于是,齐国丞相晏子就定下计谋要除掉他们三人。晏子请齐景公拿出来两个桃子送给这三个人,叫他们根据功劳大小吃桃,结果,三个人相争,最后都放弃桃子自杀了。

《世说新语》中有一个大家更为熟悉的植物典故,曹操行军迷路,士兵们饥渴难耐,不愿意前进。于是,曹操对部队宣称,前面有个大梅林,结了很多又大又甜的果实,可以解渴。士兵们听到了,口中分泌出唾液,感觉瞬间就解渴了。这也是成语"望梅止渴"的由来。

梅、兰、竹、菊是中国国画中的传统题材,在文人画家眼中,这四种并非单纯是植物的意象,还是君子的化身。梅花傲霜斗雪;兰花洁身自好;翠竹虚怀若谷;菊花淡雅高洁。画家笔下的崇山峻岭、亭台园林、花鸟鱼虫,以及人物画中服装纹饰上的植物,动物写生中搭配植物景致等。植物元素在国画中无处不在。清代画家郑板桥,就以擅长画竹子著称。

古代生活礼仪中,婚丧嫁娶也离不开植物。丧礼中的粗麻、菅草秸编制的草鞋、竹子做成的孝杖、桑木制作的孝簪以及各种木材制作的棺木。婚礼中,新婚夫妇的床上要放置枣和栗子,取早生贵子、早立子嗣的谐音。

我国传统的中医文化,更是把植物的功效发挥到极致。关于中医叫做"杏林"的由来。东晋道士葛洪著书《神仙传》,里面记载了三国时吴国神医董奉的故事。相传董奉隐居山林,每天为百姓治病,分文不取。但是如果是患重病的人被治愈,要到山上种杏树5棵,如果是一般的病被治愈,就种1棵杏树。这样过了许多年,山上的杏树已经达到了数十万棵,形成了一片茂密的森林,从此以后"杏林"就成了中医的代称。

中国古人重视农业,历代农书中专门讲述农业生产技术,经济植物的栽培方法等。比如《齐民要术》《四时纂要》《陈敷农书》《农桑辑要》等等。从晋代到清代,一些作者还编写了植物学专著,比如晋代嵇含著有《南方草木状》,宋代周师厚著有《洛阳花木记》,明代朱橚著有《救荒本草》,清代吴其濬著有《植物名实图考》等。

古人赋予了植物丰富多彩的含义。植物在中国的传统文化扮演了各种角色,我们的衣食住行,文化意识,都开不开植物。一种简单的植物名称,也能让我们浮想联翩,产生各种喜怒哀乐的情绪。

即使在我们今天居住的面积只有1.64万平方千米的北京,植物数量也达2000多种,是世界上生物多样性丰富的大都市之一。

出淤泥而不染，濯清涟而不妖

植物和人类的关系密不可分。从生态学角度分析，植物的繁盛或衰落，对人类自身乃至整个生物圈都有影响。

无论是从自然的生物、生态方面，还是从文化、地理的人文角度，植物一直都影响着人类，并为我们提供丰富的食物来源。因此，要想保护好我们的生存家园、守护好我们的精神圣地，爱护植物首当其冲。

陶渊明爱菊，王安石爱梅花，李清照喜欢白海棠。在小学语文课文里收录的宋代苏轼的《惠崇春江晚景二首》中："竹外桃花三两枝，春江水暖鸭先知。蒌蒿满地芦芽短，正是河豚欲上时"，这一首诗里，就描写了4种植物。

古代诗词中，与莲这个植物元素相关的有很多。唐代王昌龄的《采莲曲》："荷叶罗裙一色裁，芙蓉向脸两边开。乱入池中看不见，闻歌始觉有人来"；宋代周敦颐的《爱莲说》："出淤泥而不染，濯清涟而不妖"；宋代杨万里《晓出净慈寺送林子方》："毕竟西湖六月中，风光不与四时同。接天莲叶无穷碧，映日荷花别样红"。

大家是否想过《爱莲说》这篇文章中，莲出淤泥而不染的原因呢？一些科普文章认为，莲出淤泥而不染，是因为莲叶具有"自洁效应"。究竟是不是如此呢？

生命之绿 *The green of life*

在电子显微镜下,我们能观察到莲叶的上表皮是由一个个特殊的表皮细胞组成的。细胞的表面布满了高 10~20μm(微米)、宽 10~15μm 的突起。根据研究,亚洲人一根头发的直径是 80~120μm,这些特殊的表皮细胞比人类的头发丝还要纤细。

莲叶的表面还被覆着一层疏水的蜡,正是因为特殊的微米结构,使得水滴落在莲叶上时,会被突起和蜡隔离开表皮细胞,从而滚动形成"玉盆琼珠",把叶表面的脏东西黏住并带走,最终实现自洁。莲叶的自洁本领确实存在,但是,这并不能用来解释"出淤泥而不染"。刚从淤泥中钻出来莲叶的幼叶被芽鳞包裹得严严实实,根本就接触不到淤泥。只有在钻出淤泥之后,花和叶才会迅速生长。

所以,"出淤泥而不染"并不是莲叶的自洁效应,而是因为芽鳞的包裹。

可使食无肉，不可使居无竹

竹子是古代诗歌中常见的文化意象。中国最有名的竹子故事是魏晋时代的竹林七贤。这七个人指的是魏末晋初的名士：阮籍、嵇康、山涛、刘伶、阮咸、向秀、王戎。他们在生活中不拘礼法，经常在竹林中聚会欢宴。

古人将竹子与高洁的气质联系在一起，创作出许多脍炙人口的名篇。比如清代郑燮的《竹石》："咬定青山不放松，立根原在破岩中。千磨万击还坚韧，任尔东西南北风。"郑燮是扬州八怪之一，号板桥，人称板桥先生。他擅长画竹、兰、石、松、菊，特别是画竹子成就最高，他画的竹子气韵生动，挺劲孤直，具有倔强不驯之气，被世人视为郑板桥自己的人格写照。

宋代文学家苏轼说："可使食无肉，不可使居无竹"。梅、兰、竹、菊被称为四君子。竹，松，梅被称为岁寒三友。这些雅称寄托了古代士大夫们的高洁情怀。

竹子原产于中国，广泛分布在长江流域、华南和西南各地，它适应性强，几乎四季常绿。一般认为全世界的竹类有70多属1400余种。其中较为坚硬的木本竹类有近50属约1000种，还有少数竹子是草本竹类。中国共有37属约500种竹子，几乎都是木本竹类，草本竹类仅在我国台湾省发现一种。我们熟悉的大熊猫最喜欢的食物箭竹，也是木本竹类。

竹子被称为"植物钢铁"，它有着很强的抗压和抗弯能力。在汉代，竹子就被用在建筑上，能工巧匠用竹子为汉武帝建造了甘泉祠宫。在我国西南地区，有许多竹楼，它们是百姓家的房屋。

竹子是一种速生型植物，有些品种一天之内就能长高1米。它的弹性和韧性比木材大，人们利用竹子的弹性和韧性作为编织的材料，这就使竹子比木材有了更广泛的用途。因此，竹子是古人最早了解和应用的植物。

在7400多年前的湖南高庙遗址和7000多年前的浙江河姆渡遗址，都发现了用竹篾编织的席子。在距今4700多年前的浙江良渚文化钱山漾遗址中，发现了竹席、竹篓等竹器。商代晚期殷墟郭家庄墓葬中也发现了竹篓。说明在几千年前，中国南方的竹手工业已经十分成熟。

甲骨文中的"册""典"二字，就好像是把木片或竹片编成一串的样子。

汉字中，很多偏旁都是竹字头。比如，算盘的前身是"筹算"，是用竹签做筹码来进行运算的。所以，"算"和"筹"字都是竹字头。文房四宝之一的毛笔，是用竹子做成笔杆。所

生命之绿 The green of life

以,"笔"是竹字头,放毛笔的容器,是竹筒,"筒"也是竹字头。

早在两千多年前,世界上第一座农田水利灌溉工程——四川都江堰中使用了大量的竹子。世界上最古老的自来水管是用竹子制作的,所以"管"也是竹字头。

唐代刘禹锡的《陋室铭》中:"无丝竹之乱耳,无案牍之劳形",其中,"丝竹"代指乐器,"丝"是弦乐器,"竹"是管乐器。可以说,古代中国有了竹子制作的乐器,才有了音乐。考古学家在湖北随县的曾侯乙墓中发现了竹制的十三管古排箫,这是考古文物中发现年代最早的排箫。在唐代,乐器演奏者被称为"竹人",翻开字典,竹字头的乐器有很多:筝、笛、箫、竽、筘、笙、箜、簧……

第4章 中国文化中的植物
Part 4 Plants in Chinese Culture

唯有牡丹真国色,花开时节动京城

2019年,时值国庆70周年,中国花卉协会发出《投票:我心中的国花》项目,面向公众投票征集国花。在几十万选票中,牡丹以绝对优势位居第一,第二名和第三名分别是梅花和兰花。在征集中确定国花的基本条件是,起源于中国,栽培历史悠久,适应性强,分布广泛,品种资源丰富;花姿、花色美丽大气,能反映中华民族优秀传统文化和性格特征;文化底蕴深厚,为广大人民群众喜闻乐见;用途广泛,具有较高的生态、经济和社会效益。

中国是世界牡丹的发祥地和世界牡丹王国。牡丹作为历史悠久的中国特有木本名贵花卉,一直受到国人的喜爱。清末,慈禧太后下旨赐予了牡丹国花的身份。

虽然建国之后,我国没有确立国花,但是,牡丹早已成为洛阳、菏泽、铜陵、宁国市、彭州、牡丹江等市的市花,每年国庆的大花篮中,牡丹都绽放在显著的位置。

《诗经·郑风·溱洧》中,牡丹作为爱情的信物被提及。"维士与女,伊其相谑,赠之以芍药。"根据考证,春秋时期牡丹与芍药统称为芍药,二者花型相似,花期相近,同属芍药科芍药属。

魏晋时期,民间有了木芍药和草芍药的说法,木芍药即牡丹,草芍药即今天的芍药。牡丹的茎为木质,牡丹枝干木质化,冬季仅叶片脱落;芍药的茎为草质,芍药茎秆没有木

生命之绿 The green of life

质化,秋冬季落叶后地面部分枯萎。牡丹叶片宽,先端常常再裂成三瓣,正面绿色中略呈黄色;芍药叶片狭窄,先端尖而不再分裂,正反面均为深绿色。牡丹花多独朵顶生,花型大;芍药花则一朵或数朵顶生,花型比较小。牡丹通常在4月下旬开花,芍药在5月上中旬开花,故有"谷雨三朝看牡丹,立夏三朝看芍药"之说,也就是说,在自然环境下,牡丹比芍药早半个月左右开花。

牡丹进入植物药物学记载于汉代的《神农本草经》,明代李时珍在《本草纲目》中描述"牡丹虽结籽而根上生苗,故谓'牡'(意思是可以无性繁殖),其花红故谓'丹'。"从药用角度认识牡丹,迄今已有2000多年的历史。

牡丹作为观赏植物栽培,始于南北朝。隋朝时期有了国家牡丹园,牡丹人工栽培在我国历经了1600多年历史。清代还普及了牡丹种植的熏花技术,和现在的温室大棚技术类似。

唐代,牡丹栽植中心移至长安,人们对牡丹的热爱达到高峰。唐代诗人刘禹锡的《赏牡丹》:"庭前芍药妖无格,池上芙蕖净少情。唯有牡丹真国色,花开时节动京城。"说明牡丹已经成了家喻户晓的名花。唐代诗人白居易《秦中吟十首·买花》节选中:"有一田舍翁,偶来买花处。低头独长叹,此叹无人喻:一丛深色花,十户中人赋!"说一丛红牡丹的价钱,竟然抵得上十户中等人家一年的税赋。说明当时的达官贵人们追捧名贵的牡丹花,不惜重金购买。

唐代李正封《牡丹诗》:"国色朝酣酒,天香夜染衣",这也是成语"国色天香"的由来,从此,人们用"国色天香"来形容牡丹之美。

宋代周敦颐《爱莲说》中称牡丹为"花之富贵者也",说明了在唐宋时期,社会赋予牡丹的美好寓意。数千年来,牡丹的形象在中国被广泛应用在诗歌、绘画、刺绣、印染、雕刻等。

宋代还出现了牡丹的植物学专著,如欧阳修的《洛阳牡丹记》、陆游的《天彭牡丹谱》等。

虽然牡丹起源于中国,但是,关于牡丹在植物学被记录和命名的时间却并不久远。牡丹在植物分类学上,属于芍药科芍药属。全世界芍药属共有34种,芍药属又分为牡丹组和芍药组。其中,牡丹组全部为中国特有。

直到1804年,英国植物学家和植物绘画巨匠亨利·查尔斯·安德鲁斯第一次为牡丹进行了描述和分类。

第一位描述牡丹野生种类的人是法国植物学家弗朗西(A.R.Franchet),他于1886年发表了新种,即产自我国西南地区特有的滇牡丹(Paeonia dclavavi Franch)。

直到1958年,中国植物分类学家、教育家方文培教授对芍药属作了全面的研究,成为第一位为中国牡丹组分类的中国人,他在牡丹组中记录了6个种。

近年来,我国著名植物学家洪德元院士及其团队又对中国特有种类芍药属牡丹组植物作了更深入系统的基因研究,认为我国芍药组有7种2亚种,牡丹组有9个野生种和一个栽培杂交种。

目前,中国牡丹属的野生种表现出了令人担忧的等级状态,野生中原牡丹属于灭绝植物。许多野生牡丹品种都属于易危植物和濒危植物。

但是,国人热爱牡丹千年未改。每年春季4~5月,许多人从全国各地赶赴河南洛阳和开封等地,观看牡丹怒放的美景。

牡丹作为中国特有植物,汇集了丰富的遗传资源。牡丹在自然状态下不同种,不相遇,因而无法形成杂交种。而人们在进行园中移栽时,将不同野生牡丹种在一起,借助蜂等媒介发生了自然杂交,产生了千姿百态的牡丹花品种,使牡丹成为当之无愧的花中之王,也被外国人誉为"King of Flowers"。

唯有此花开不厌,一年长占四时春

如果要选一种花代表爱情,那一定是玫瑰。玫瑰花作为爱的信物,从近代起就受到人们的热爱。但是,花店里售卖的大多数玫瑰,很多都是月季。

根据《中国植物志》的分类:玫瑰、月季和蔷薇都隶属于蔷薇科、蔷薇属、蔷薇亚属。

玫瑰在西方与月季通常都被称为"rose",因为杂交玫瑰是由蔷薇属下的各物种杂交选育产生的,所以玫瑰是多种蔷薇属植物的统称。

汉语博大精深,中国人用不同的名字来称呼蔷薇属的植物,比如把西方月季翻译成玫瑰,把传统月季称为月季。

南宋诗人杨万里的《红玫瑰》:"非关月季姓名同,不与蔷薇谱牒通,接叶连枝千万绿,一花两色浅深红。风流各自胭脂格,雨露何私造化工?别有国香收不得,诗人熏入水沉中。"既描述了玫瑰与月季、蔷薇的不同,又描述了玫瑰的枝叶形态和颜色变化。

从植物特征上,月季和玫瑰差别不大。玫瑰茎部密生锐刺,一枝只开一朵花,花瓣直立,叶脉深凹褶皱;而月季刺疏,标准的花形是"满芯、翘角",往往数朵同生一茎,叶片光大鲜亮。

因为玫瑰的特点是叶片发皱,所以它的拉丁名是 *Rose rugosa* Thunb.,种加词 rugosa 的意思是多皱的,描述的就是玫瑰的叶片。

月季的拉丁名是 *Rosa chinensis* Jacq.,从名称上我们就可以看出,月季的原产地在中国,中国月季也被称为"世界月季之母"。"月季"这一名称的由来,是因为它一年四季不分季节几乎每个月都能开花,所以又叫"月月红"、"长春花"。

考古发现,月季是华夏先民北方系,也就是传说中黄帝部族的图腾植物。它是中国十大名花之一,被誉为"花中皇后"。月季早在汉代就开始栽培,唐宋以后达到繁盛。历代文人留下了很多赞美月季的诗句。如宋代文豪苏轼描写月季四季常开的诗句:"花落花开无间断,春来春去不相关。"

国人喜爱月季,它是中国首都北京市、天津市、石家庄、郑州、南昌、大连、锦州、青岛等 53 个市的市花。

18 世纪时,欧洲人将四季开花的中国月季引入西方,与玫瑰和蔷薇杂交。经过反复杂交和定向选育,培育了许多杂交新品种,玫瑰的种植很快风行世界,成为所有花卉中最著名和最受欢迎的种类。

法国皇帝拿破仑的妻子,皇后约瑟芬在巴黎市郊建立玛尔梅森花园,这是当时世界上规模最大的玫瑰园。约瑟芬请著名法国花卉图谱画家皮埃尔·约瑟夫·雷杜德为她的玫瑰园绘制了世界上第一份《玫瑰图谱》,这份图谱耗时 20 年,收录了大约 170 个品种,被后人誉为"玫瑰圣经"。

在中国,早期的玫瑰并不是花名,而是一种宝石的名字。"玫瑰"是玉字旁,汉语中一

生命之绿 *The green of life*

般"玉"在做左偏旁时,会失去一点,就好像"王"字。《红楼梦》中,贾宝玉这一辈份的人,名字都是从"玉"做偏旁,除了贾宝玉之外,其他堂兄弟叫:贾琏、贾琮、贾珍、贾珠、贾环等,就是"玉"在做左偏旁时,会失去一点。

《诗经·渭阳》曰:"何以赠之?琼瑰玉佩。"这里"瑰"就是指"美石"。东汉许慎的《说文解字》中,"玫"作为名词是"火齐(ji)珠"或"美石"的意思,"瑰"和"玫"同义。《韩非子》一书首次把"玫瑰"作为词语组合使用。古人形容"玫瑰"的宝石颜色,是纯铜一般的紫红色,这种矿石可以呈现很多薄层。今天的矿物学家认为玫瑰就是锂云母。

中国古人很早就将蔷薇属植物分为至少7类品种:月季、玫瑰、蔷薇、缫丝花、木香、金樱子、荼蘼。

玫瑰作为植物的栽培历史可以追溯到2000年前的汉代。晋代葛洪《西京杂记》中描述的"玫瑰树",就是汉武帝时期栽培的。中国玫瑰之乡平阴,早在唐代开始种植玫瑰。玫瑰颜色紫红,就好像宝石玫瑰的颜色,并且气味芳香,能制作各种香料,一些品种还能食用,比如我们常吃的云南鲜花饼,主要采用中国特有的云南"土著"玫瑰—滇红玫瑰制作的,这是一种可食用重瓣红玫瑰。

因为在文化上的美好寓意,玫瑰被广泛认为是爱和美丽的象征。世界上许多国家都用玫瑰作为国花。玫瑰是美国、英国等14个国家的国花,也是美国爱荷华州、北达科他州、乔治亚州和纽约州四个州的州花,是加拿大艾伯塔省的省花。

在中国,玫瑰是沈阳市、拉萨市、兰州市、银川市、乌鲁木齐市、济南市等十多个市的市花。